Hydraulics in Water and Waste-Water Treatment Technology

Hydraulics in Water and Waste-Water Treatment Technology

by

Imre Horváth C. Sc.

Associated Professor
University of Horticulture
and Food Industry, Budapest, Hungary

AKADÉMIAI KIADÓ

Budapest

JOHN WILEY & SONS

Chichester · New York · Brisbane · Toronto · Singapore

This book is the revised version of the original Hungarian
„A csatornázás és szennyvízkezelés hidraulikája",
VIZDOK, Budapest

Translated by

G. Jolánkay

Published by

Akadémiai Kiadó
Prielle Kornélia u. 19–35
H-1117 Budapest, Hungary

and

John Wiley & Sons Ltd,
Baffins Lane, Chichester,
West Sussex PO19 1UD, England

British Library Cataloguing-in-Publication Data

A catalogue record for this book is available from the British Library

ISBN 0 471 94422 X

ISBN 963 05 6478 5 (Akadémiai Kiadó)
ISBN 0 471 94422 X (Wiley)

Printed in Hungary by Akadémiai Kiadó és Nyomda Vállalat, Budapest

Contents

Preface

In assembling the subjects of this book, my objective was to provide a review for engineers dealing with the hydraulics of water and waste-water treatment technologies.

The book consists of 13 main chapters in which the main emphasis is laid on technological processes of higher practical significance. No full coverage of the subject can be claimed due, among other things, to limitations on the size of the book, although available research results might have allowed for more detailed discussion. Nevertheless, it is my strong belief that this work will provide a useful supplement to publications that have previously appeared in this field.

The background of this book was its Hungarian version, previously published in two volumes. The first of these volumes provided the basis for the present English version (the second volume contained only calculation examples). The widespread interest shown by Hungarian and foreign experts in the Hungarian book has been a great satisfaction to me and it has been confirmed by the fact that copies of this first edition are no longer available.

I release this book for publication in the hope that it will provide modest but useful additional information for experts working in this field and, last but not least, that it might perhaps be used as a teaching aid in the university and postgraduate training of specialists.

Imre Horváth

I. Mechanical treatment

1 Fine and rough screens (trash racks)

1.1 On waste-water screens, in general

The purpose of waste-water screens is to remove larger floating and suspended debris and sediments from the sewage, thus facilitating the conveyance, lifting and treatment of waste waters and preventing clogging. Screens are essential components of the mechanical treatment system. Screens should be installed in pumping stations in front of the pumps and in the sand traps and preliminary settling basins of waste-water treatment plants. Waste-water screens are seldom applied alone. Such an exceptional case is, however, that of effluent outfall into rivers with high discharge rates or into seas, when the major objective is to retain rough-floating and suspended debris contained in the waste water.

Depending on the spacing of the screen rods, rough screens (trash racks) and fine screens can be distinguished. Rod-spacing in rough screens is generally 50 mm, but might be as much as 150 mm in special cases. In the case of manual cleaning the rod-spacing is 60—80 mm, while for mechanical cleaning 40—60 mm is desirable. Larger rod-spacing permits the passage of paper, smaller debris and mud, the removal of which can be performed in the preliminary settling tank. Rough screens will, however, remove larger coarse litter and debris. Rough screens are mostly designed for combined sewer systems. The rod-spacing of fine screens is 10—50 mm. Generally, 20 mm should be considered as an initial design value.

Screens are installed at a certain angle of inclination. When sufficient space is available, the angle of the screen to the horizontal should be 20—30°. In order to increase the efficiency of space utilization, the screen angle might be increased to 60—75°. Screens should be installed in screen-shafts.

1.2 Hydraulic design

The principle of designing sewage screens hydraulically relies on consideration of the so-called flow-resistance. Namely, screens cause local head losses and the head loss h_v can be calculated by the Weissbach equation, known from the relevant literature:

$$h_v = \zeta \frac{v^2}{2g},$$

$$(1.1)$$

where ζ is the loss coefficient; v is the mean flow velocity on arriving at the screen; g is the acceleration of gravity.

Head loss h_v expresses the screen's resistance to flow and it appears in the form of a backwater rise, that is to say, in a water-level difference $h_v = \Delta h$.

It is to be noted that the velocity head in eq (1.1) refers to the cross-sectional mean velocity in front of the screen. The reason why this should be emphasized is that in hydraulic design local head losses are generally calculated — with a few exceptions — on the basis of cross-section parameters downstream (down-gradient) of the structure causing the loss.

Head-loss coefficient ζ involves all parameters and variables that are defined by the geometry and the manner of installation of the screen. For the determination of the value of ζ, various numerical and graphical guides have been developed on the basis of experimental results; some of the more important ones will be presented below.

The Kirschmer relationship

In the field of waste-water technology, the most widely used screen-design method is based on the relationship published in 1926 by Kirschmer, in the form of

$$h_v = \beta \left(\frac{d_p}{k_p}\right)^{4/3} \sin \alpha \frac{v^2}{2g}, \tag{1.2}$$

where β is a shape-coefficient depending on the geometric design of the screen-rod; d_p is the width of the rods (measured perpendicularly to the direction of flow); k_p is the rod-spacing, and α is the angle between the plane of the screen and the horizontal plane.

Table 1.1. Shape coefficients of screen rods (after Kirschmer 1926)

Shape of rod	a, h, i, j, k, l	b	c	d	e	f	g
β	2.34	1.77	1.6	1.0	0.87	0.71	1.73

Remark: The data of the above Table and of Fig. 1.1 differ from each other, since they refer to two different series of experiments. Consideration of the data of Fig. 1.1 is recommended for design purposes.

Table 1.1 presents some basic data for the shape-coefficients, considering rod dimensions as shown in Fig. 1.1.

The Kirschmer equation is valid — at different angles of inclination — for flows perpendicular to the screen.

Spanglers' investigations

The experiments carried out by Spangler (in Wechmann 1966) supported the validity of eq (1.1) for the hydraulic design of screens and trash-racks. The results

$$h_v = \beta \left(\frac{d}{k}\right)^{\frac{4}{3}} \sin \alpha \; \frac{v^2}{2g}$$

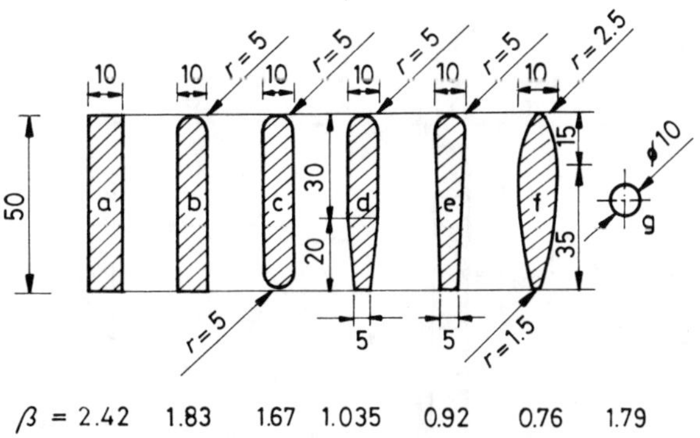

$\beta = 2.42 \quad 1.83 \quad 1.67 \quad 1.035 \quad 0.92 \quad 0.76 \quad 1.79$

Fig. 1.1. Different screen rod profiles (after Kirschmer 1926)

relate to the determination of loss coefficient ζ, for different screen-rod dimensions. ζ values for the relative throughflow cross-section

$$\varphi = \frac{b}{s+b} = 0.63$$

are given in *Table 1.2* for the cross-section design alternatives of *Fig. 1.2*. From the practical point of view, consideration of inflow-direction angles, other than zero $-\alpha^* \neq 0°$, is important. (In the above equation $b=k_p$ and $s=d_p$.)

Table 1.2. Loss coefficients of screens and trash racks
(after Spangler, in Wichmann 1966)

Shape of rod	Angle of inflow direction α^*			
	0°	30°	45°	60°
a	1.13	1.46	2.05	4.26
b	0.86	0.76	1.29	2.45
c	0.78	0.71	1.29	2.81
d	0.48	0.43	0.94	1.19
e	0.42	0.68	1.29	3.05
f	0.38	0.22	0.67	1.84
h	1.13	1.88	2.75	5.15
i	1.13	1.81	2.72	4.26
k	—	1.53	2.32	3.43
l	1.13	1.62	2.12	3.88

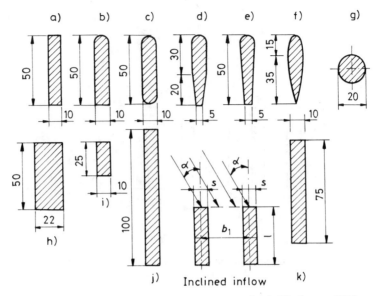

Fig. 1.2. Different screen rod profiles (after Spangler, in Wechmann 1966)

The relationship of Berezhinsky

The Berezhinsky equation

$$\zeta = K_r \left(\frac{s}{s+b}\right)^{1.6} \sin \alpha f\left(\frac{l}{b}\right) \qquad (1.3)$$

serves for the determination of the value of loss coefficient ζ of screens and trash-racks, where K_r is a shape coefficient depending on the design of rods and l is the longer dimension of the rod cross-section. The above relationship can be applied in the perpendicular inflow direction only. *Figure 1.3* shows a design-guide graph for

a) screens of square-profile rod with $\alpha = 70°$ of inclination;
b) for profiles other than square;
c) for inclinations other than 70°.

Starting with coefficient ζ_1 of case *a* and applying correction coefficients φ and k_α for cases *b*, and *c*, the value of loss coefficient ζ_φ for a general case is obtained as

$$\zeta_\varphi = \varphi k_\alpha \zeta_1 . \qquad (1.4)$$

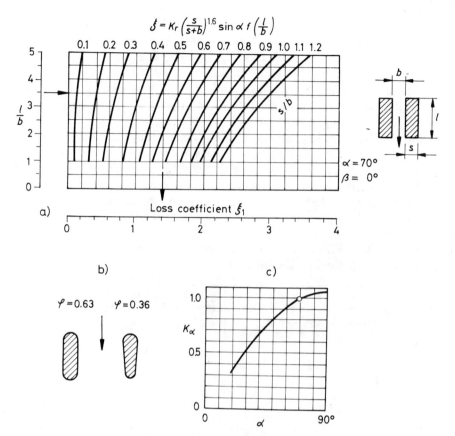

Fig. 1.3. Nomogram for calculating the loss coefficients of screens and trash racks (after Berezhinsky, in Wechmann 1966)

a) For screen rods with square profile placed at 70°

b) Correction factor for profiles other than square

c) Correction factor for screen inclinations other than 70°

1.3 Design principles

Making use of the above equations and graphical guides, the hydraulic design can be carried out. In sewage-technological practice, the most frequently needed *design* parameters are the rod-spacing or the rod-width. The determination of head loss h_v is the most frequent task when *control* is the objective of the calculation. Consideration of initial experimental values is also needed for the calculations.

One of the basic principles is that the flow velocity between the rods must not exceed 1.0 m/s, since in such a case the flow might easily carry away the screens. The desirable velocity range is 0.5—0.8 m/s. Nevertheless, care should be taken to ensure flow velocities within the above limits in the access and outflow canals also, so as to prevent the settling of mineral particles (i. e., sand) in these sections. The backwater effect of the screen should be taken into consideration as well. A head increase of about 5.0 cm is generally considered as a backwater effect.

According to certain regulations, a safety value should be added to head loss h_v, as obtained from the formulas. This provides, for example, a means of considering the condition that screens in operation are always contaminated to a certain extent. Similarly, inflow directions other than perpendicular can be taken into account in this way. Screens subject to less frequent cleaning can be considered with an additional head-loss value of a maximum of 10 cm, while that for frequently cleaned screens should be 5.0 cm.

The approximate values of contraction coefficient ψ, required in considering the effect of accumulated screenings, are as follows:

	In dry weather	In wet weather
Manually cleaned screens	0.60	0.75
Mechanically cleaned screens	0.75	0.85

2 Principles of mechanical cleaning by gravitation

Settling is one of the basic classical operations in water and waste-water treatment technologies. According to the investigations of Camp (1946), the investment costs of settling facilities make up about one third of the total cost of constructing treatment plants. This fact in itself justifies the substantial concern devoted, on a global scale, to the design aspects of settling facilities, and therein especially to the flow conditions. The larger size of the chapters in this work devoted to the hydraulics of settling technologies is also justified thereby. In this chapter, the theoretical principles of gravitational settling are discussed, with distinction being made between spherical and non-spherical particles, discussion of the free and restricted settling conditions of floc-like particles and also consideration of stagnant and flowing media.

2.1 The paradox of d'Alembert and the drag resistance

According to the paradox of d'Alembert, the resultant of hydraulic forces acting on a finite body immersed in an ideal liquid of infinite extent is probably zero, if the flow around the body is free from sources and vortices. A special case of this is the flow around a spherical body. According to the above paradox, the resultant force of the dynamic forces acting on a sphere immersed in an ideal liquid is zero, when uniform linear flow is assumed. In this situation, the emphasis is laid on the *relative movement*. Finally, it can be stated that without external forces the spherical body immersed in an ideal infinite liquid maintains uniform and straight movement if the liquid is stagnant or moves uniformly along straight flow lines. The same conclusion can also be obtained by the following — theoretically derivable — expression:

$$P = \frac{2}{3} r^3 \pi \varrho \frac{\partial w_t}{\partial t} = \frac{V_{sphere}}{2} \varrho \frac{\partial w_t}{\partial t}, \qquad (2.1)$$

where P is the accelerating force; r is the radius of the spherical body; V_{sphere} is the volume of the spherical body; w_t is the relative velocity of the body that changes in time; t is the time, and Q is the density of the liquid.

Consequently if $\partial w_t/\partial t = 0$ then $P = 0$.

2*

In reality, settling in the settling structures of treatment operations takes place in viscous liquids. Consequently, there is a drag resistance (a resistance to flow) occurring as the resultant force of *friction* forces and *shape* and *pressure* resistances (i.e., that of the forces of friction and inertia). The force of inertia can also be considered as a collision force. The resistance due to shape and pressure originates from the vortices that occur behind (downstream of) the body. This means that the pressure behind the body encircled by the flow will be less than that which would occur in ideal liquids. At lower velocities — as proven by Stokes — shear forces play the decisive role, while inertia becomes negligible. At higher velocities, however, the force of inertia will be the dominating one.

2.2 The basic equation of settling dynamics

The dynamic equation of the movement of a settling body, placed in a liquid, can be described on the basis of the projections of the forces acting in the direction of the movement. Generally, the following forces are acting: the force of gravity G, the buoyant force A; the drag force (the drag resistance) W; and the force of inertia $P = ma$. The projection equation of the above acting forces is obtained as

$$G - A \pm W = P \tag{2.2a}$$

For spherical bodies and by taking the squared-resistance law of Newton into account, the following equation can be obtained;

$$\frac{d^3 \pi}{6}(\varrho_1 - \varrho) \pm C_w \varrho \frac{d^2 \pi}{8} w_t^2 = \frac{d^3 \pi}{6g}\left(\varrho_1 g + \frac{\varrho g}{2}\right)\frac{dw_t}{dt}, \tag{2.2b}$$

where d is the diameter of the spherical body, C_w is the coefficient of resistance (drag); ϱ_1 and ϱ are the densities of the settling particle and the liquid, respectively.

On the basis of the above equation, the following basic cases can be distinguished:

a) In a general case — for accelerating movement — all terms of the equations have their role.
b) At low initial velocities, W becomes negligible and thus $G - A \approx P$.
c) For uniform movement, $P = 0$ and thus $G - A \pm W = 0$.
d) For downward movement, the sign of W is negative, while it is positive for uplifting particles.
e) For suspended particles, $w_t = 0$ and thus $G = A$.

In the subsequent sections of this chapter the cases under items a) and b) will be dealt with, in conjunction with the analysis of the accelerating phase of settling. The main features of the settling operations will be discussed in relation to the cases under items c) and d).

2.3 The accelerating phase of settling

The extent of initial acceleration

The initial phase of the settling movement of a particle placed in a liquid medium is characterized by acceleration. Acceleration occurs due to the force of gravity. The drag resistance increases simultaneously. A uniform movement with a velocity $w_t = w$ will take place once the balance of the forces acting is achieved. In practice, this velocity w is termed the *final settling velocity*, or for short the *settling velocity*. The initial accelerating phase — as will be revealed later — is very short and thus can be neglected when designing settling technologies. For example, for particles of 0.05 mm diameter the uniform settling movement occurs after an initial acceleration phase of only 0.003 seconds. In special cases, however (for example when large particles are settling and especially when the settling depth is small), the phase of the acceleration should be taken into consideration as well.

For spherical particles, the initial accelerating phase of settling is expressed by eq (2.2b). At the initiation of settling, w is approximately zero and thus the *initial acceleration* is obtained as

$$\frac{dw_t}{dt} = g\, \frac{\varrho_1 - \varrho}{\varrho_1 + \dfrac{\varrho}{2}}, \tag{2.3}$$

a relationship that is independent of the size of the particles.

Time of acceleration

Theoretically, the final velocity $w_t = w_\infty$ would occur after an infinite time. For practical reasons, it is expedient to introduce t_{99}, the time for achieving 99% of the final settling velocity; this value can be determined from the equation of settling movement.

Let us assume, as the *first alternative*, that drag resistance is proportional to velocity w_t:

$$mg_A - k_{St} w_t = m\frac{dw_t}{dt}, \tag{2.4}$$

where m is the mass of the moving body; $g_A = g(\varrho_1 - \varrho)/\varrho$ is the so-called *acceleration of Archimedes*; ϱ_1 and ϱ are the densities of the settling particle and the medium, respectively; and k_{St} is the Stokes' constant.

A solution of the differential equation

$$w_t = w_\infty \left[1 - \exp\left(\frac{-k_{St}}{m} t\right)\right] \tag{2.5}$$

is obtained, and then

$$t = \frac{m}{k_{St}} \ln \frac{1}{1 - \frac{w_t}{w_\infty}} .$$

(2.6)

Making use of the 99% velocity concept introduced above $w_t = w_{99} = 0.99 \, w_\infty$, and

$$k_{St} w_\infty = mg_A; \qquad w_\infty = \frac{mg_A}{k_{St}},$$

(2.7)

therefore

$$t_{99} = \frac{w_\infty}{g_A} \ln \frac{1}{1.0 - 0.99} \approx \frac{w_\infty}{200} .$$

(2.8)

For the *second alternative* let us assume that the drag resistance is proportional to the square of the velocity w_t:

$$mg_A - k_{Ne} w_t^2 = m \frac{dw_t}{d_t},$$

(2.9)

where k_{Ne} is Newton's coefficient.

Solving the differential equation, it is obtained that

$$w_t = w_\infty \, \text{th} \left(\frac{k_{Ne}}{m} t \right)$$

(2.10)

and further

$$t = \frac{w_\infty}{2g_A} \ln \frac{1 + \frac{w_t}{w_\infty}}{1 - \frac{w_t}{w_\infty}},$$

(2.11)

and finally

$$t_{99} = \frac{w_\infty}{2g_A} \ln \frac{1 + 0.99}{1 - 0.99} \approx \frac{w_\infty}{400} .$$

(2.12)

On the basis of the above relationships, it can be readily concluded that velocity w_t of the settling particles approaches the final velocity $w = w_\infty$ within a very short period of time. Finally, it is to be noted that the two alternatives discussed above correspond to the linear and square resistance laws of Stokes and Newton, respectively, that will be discussed in more detail subsequently.

2.4 Newton's general relationship

As was already mentioned above, the classical theoretical basis of calculating the settling velocity is related to the introduction of the notion of the drag resistance. Newton recognized the *squared relationship* between velocity and the drag force, a law that was named after him. The way of thinking that resulted in Newton's theorem can be described as follows: Assume that the projection area, perpendicular to the direction of motion of a settling particle that moves with velocity w in a medium of density, is A. This movement displaces a mass of $m = \varrho A w$ during each unit period of time with a velocity of w_*. In the first step $w_* = \text{const } w$ can be assumed. According to the theorem of momentum, the momentum taken up during a unit period of time, or equivalently (according to the theorem of action and reaction) the resistance force acting on the frontal area of the moving body, is:

$$W = mv_* = \varrho A w w_* = \text{const } \varrho \, A w^2. \tag{2.13}$$

The above relationship is Newton's resistance law, discovered in 1729, and — in accordance with the above-presented original way of thinking — it considers the *force of inertia* (the collision) only. The effects of friction were subjected to research only at a later stage of the respective field of science. According to eq (2.13) the drag resistance force W is proportional to the square of the velocity (the squared resistance law). Later it was demonstrated that the value of const is not constant at all, but varies as functions of the Reynolds number, corresponding to the moving body, and also as of a shape coefficient ζ. Finally, the drag resistance force is described as

$$W = C_w \varrho A \frac{w^2}{2}, \tag{2.14}$$

where $C_w = C_w(\text{Re}; \zeta)$.

For a spherical body moving along a straight line with uniform speed, one obtains that

$$W = C_w \frac{d^2 \pi}{4} \frac{\varrho w^2}{2} = C_w \varrho \frac{d^2 \pi w^2}{8} = C_w r^2 \pi \frac{\varrho w^2}{2}. \tag{2.15}$$

The term $\varrho w^2/2$ is the so-called dynamic pressure, while the impact pressure acting on the area $r^2 \pi$ is $r^2 \pi \varrho w^2/2$. Resistance coefficient C_w, as a dimensionless unit, can thus be defined for spherical bodies as

$$C_w = \frac{W_{\text{sphere}}}{r^2 \pi \varrho \, w^2/2}. \tag{2.16}$$

Newton's relationship can also be obtained by dimension analysis. Starting with the general relationship of $W = f(w, d, \varrho, \eta)$, the dimension equation $[W] = [w^x d^y \varrho^p \eta^q]$ can be written. Using the basic units of M (mass), L (length) and T (time) one obtains that: $[MLT^{-2}] = [M^{p+q} L^{x+y-3p-q} T^{-x-q}]$, from which x, y, p and q can be derived. The relationship finally obtained will be $W = w^2 d^2 \varrho \, f(wd\varrho/\eta) = w^2 d^2 \varrho \, f(\text{Re})$, similarly as with the former method.

Following the above reasoning, the dynamic balance equation for a settling particle of density ϱ_1 and volume V can be formulated by making use of gravity force G, buoyant force A and drag force W as $G - A = W$, that is

$$V g (\varrho_1 - \varrho) = C_w A \varrho \frac{w^2}{2}; \tag{2.17a}$$

for spherical particles this becomes:

$$\frac{d^3 \pi}{6} g(\varrho_1 - \varrho) = C_w \frac{d^2 \pi}{4} \varrho \frac{w^2}{2}, \qquad (2.17b)$$

from which the settling velocity is

$$w = \sqrt{\frac{2g}{C_w} \frac{V}{A} \frac{\varrho_1 - \varrho}{\varrho}}; \qquad (2.18a)$$

further, for a particle of diameter d it becomes

$$w = \sqrt{\frac{4}{3} \frac{gd}{C_w} \frac{\varrho_1 - \varrho}{\varrho}} = \sqrt{\frac{4}{3} \frac{gd}{C_w} \frac{\gamma_1 - \gamma}{\gamma}}. \qquad (2.18b)$$

   The above relationship is termed Newton's equation for the settling velocity. Its range of validity is defined by the balance equation (2.17). According to this, the equation (2.18) is valid for a wide range if $C_w = f(Re; \zeta)$ is known. Obviously, for making actual calculations, the value of C_w must be known and in this respect results have been published by several authors. *Figure 2.1*, showing the $C_w = f(Re; \zeta)$ relationship, is presented as an example. Analogously to pipe flow, the curve presented can be characterized with *laminar, transitional* and *turbulent ranges*. In the relationships to be presented subsequently, some semi-empirical and empirical approaches to

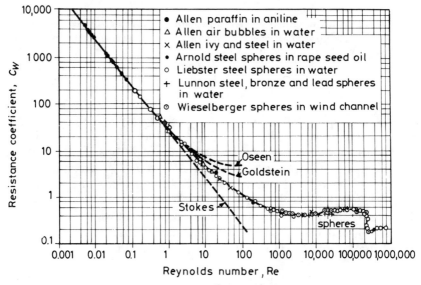

Fig. 2.1. Relationship between resistance coefficient C_w and the Reynolds number (after Camp 1953)

the general relationship $C_w = f(\text{Re}; \zeta)$ are discussed, for various ranges of the Re number. (The role of parameter ζ will be discussed in Section 2.6.)

The main *advantage* of Newton's law on settling is its general applicability. Resistance relationships for various Re number ranges can thus be discussed on a uniform basis. However, its *disadvantage* is the need for determining the value of C_w. Since the value of w is not known in advance, the first estimated values of w or Re should be substituted, and the task can then be solved by gradual approximation.

In addition to the method of gradual approximation, there are some practical *graphical methods* for estimating the corresponding values of C_w and Re directly. In this context, some examples from the literature will be presented below. From eq (2.18b) resistance coefficient C_w can be expressed as

$$C_w = \frac{4}{3} gd \frac{\gamma_1 - \gamma}{\gamma} \frac{1}{w^2}.$$

In logarithmic form this becomes:

$$\log C_w = \log \frac{4gd(\gamma_1 - \gamma)}{3\gamma} - 2 \log w;$$

the term $\log w$ can also be expressed in terms of the Reynolds number as

$$\text{Re} = \frac{wd}{v} = \frac{wd}{\eta/\varrho}$$

$$\log w = \log \text{Re} - \log \frac{d\varrho}{\eta},$$

where η and v are the dynamic and kinematic viscosities of the medium, respectively. Eliminating $\log w$ from the above two equations, one obtains that

$$\log C_w = -2 \log \text{Re} + \log \frac{4gd^3(\varrho_1 - \varrho)\varrho}{3\eta^2}, \qquad (2.19)$$

which yields a relationship between C_w and Re in the form of a straight line in the log—log coordinate system of slope $\tan \alpha = -2$. If the relationship $C_w = f(\text{Re})$ of the given settling particle is known — on the basis of *Fig. 2.1* — then the intersection of the curve and the straight line yields the solution for the values of C_w and Re. In the possession of these values, the value of w can be calculated directly.

According to Kármán's procedure, eq (2.18b) should be rearranged as

$$C_w \text{Re}^2 = \frac{4}{3} \frac{d^3 g(\varrho_1 - \varrho)\varrho}{\eta^2}. \qquad (2.20)$$

Since the term on the right-hand side of this dimensionless equation can be calculated without knowing the value of w, the term $C_w \text{Re}^2$ can be obtained numerically. At the same time, the relationship $C_w \text{Re}^2 = f(\text{Re})$ is known as a

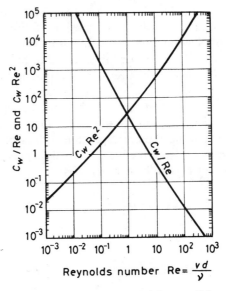

Fig. 2.2. Modified curves for drag coefficient C_w

graphical aid (*Fig. 2.2*) from which the respective values of Re and thus of w, can be obtained.

Another application example can be presented by calculating the diameter of settling particles. In this case, the basic eq (2.18b) is expediently rearranged in the following form:

$$\frac{C_w}{Re} = \frac{4g(\varrho_1 - \varrho)\eta}{3\varrho^2 w^3}.$$
(2.21)

Calculating the value of the term on the right-hand side of the above equation on the basis of data available, the Reynolds number can thus be obtained from the $C_w/Re = f(Re)$ curve of *Fig. 2.2*, which latter forms the basis of calculating the d value in question.

2.5 Settling of spherical particles

The relationships obtained for the settling of various particles in liquid media can be utilized by given conditions of validity. The characterization of the settling of spherical particles is one of the most widely discussed cases, and creates one of the classical problems of the hydraulics of settling.

2.5.1 The Stokes relationship

In the low value ranges of the Reynolds number, the linear resistance law of Stokes is valid, as mentioned above, for spherical particles. The equation is obtained as a solution for a special case of the Navier–Stokes differential equation:

$$W = 6\pi r\eta w = 3\pi dv\varrho w . \tag{2.22}$$

It can be demonstrated that two thirds of the above total resistance consists of shear forces, while one third is due to dynamic pressure. The effect of inertia does not play a role.

The dynamic equation written for the settling particle then becomes:

$$\frac{d^3 \pi}{6} g(\varrho_1 - \varrho) = 3\pi dv\varrho w , \tag{2.23}$$

whence the settling velocity is:

$$w = \frac{gd^2}{18v} \frac{\varrho_1 - \varrho}{\varrho} = \frac{gd^2}{18\eta}(\varrho_1 - \varrho) = \frac{2}{9}\frac{gr^2}{\eta}(\varrho_1 - \varrho) . \tag{2.24}$$

The above equation is the well-known Stokes relationship for settling velocity, that was published by Stokes in 1845. The *validity conditions* eq (2.24) are as follows: the settling particle is spherical, solid and has a smooth surface; its shape and dimensions do not change during settling; the surrounding liquidous medium is homogeneous and the space it fills is infinite; the effects of Brownian motion are negligible; the drag force is determined by eq (2.22); the motion is uniform and follows a straight line.

Stokes' settling law and Newton's law can be related to each other. Assuming the equivalence of eqs (2.15) and (2.22) for drag force W, the resistance coefficient C_w can be expressed as

$$C_w = \frac{12\eta}{\varrho\,rw} = \frac{24}{\dfrac{wd}{v}} = \frac{24}{Re} . \tag{2.25}$$

Eventually, identical results can be obtained also on the basis of eq (2.16). Substituting eq (2.25) into eq (2.18b), one obtains Stokes' law directly.

In order to facilitate practical calculations, the validity constraints of Stokes' law should be specified. It was experimentally proven that the upper limit can be fixed at $Re = 1.0$, although — due to analytical considerations — the upper limit of the range of validity is characterized as $Re < 0.5$. If $Re = 1.0$, then an about 10% deviation from the calculated value of Stokes' W can be expected. Namely, at higher Re numbers inertia becomes increasingly significant, a feature that had not been taken into account when deriving eq (2.22). In the case of particles settling in water, a maximum particle diameter range of 50—80 μ (0.05—0.08 mm) can be considered as design values. The lower limit of validity is affected by the Brownian motion. The lower limit, in the case of particles settling in liquids, falls into the range of colloids. The maximum particle size can be considered as 0.1—0.5 μ. According to Fair et al. (1968), the validity of eq (2.25) in calculating C_w can be expressed, for applications in sewage-treatment technology, as

$$10^{-4} \leq \mathrm{Re} \leq 0.5;$$

according to Eckenfelder (1924), the respective limits are

$$10^{-4} \leq \mathrm{Re} \leq 2 .$$

In considering the critical Reynolds number at the transition between laminar and turbulent flow, w_{crit} and d_{crit} can also be determined. Assuming $\mathrm{Re}_{crit} = 1$, the respective expressions can be obtained from Stokes' equation as

$$w_{crit} = \frac{v}{d_{crit}} = \frac{\eta}{d_{crit}\varrho}; \tag{2.26a}$$

$$d_{crit} = \left[\frac{18\eta^2}{g(\varrho_1 - \varrho)\varrho} \right]^{1/3} . \tag{2.26b}$$

If for example:

$\varrho_1 = 1.5$ g/cm³, then $d_{crit} = 1.5 \times 10^{-2}$ cm and $w_{crit} = 0.7$ cm/s;

$\varrho_1 = 2.5$ g/cm³, then $d_{crit} = 1.0 \times 10^{-2}$ cm and $w_{crit} = 1.0$ cm/s;

$\varrho_1 = 10.0$ g/cm³, then $d_{crit} = 0.6 \times 10^{-2}$ cm and $w_{crit} = 1.7$ cm/s.

In the case of air bubbles rising in water

$\varrho_1 = 0.0013$ g/cm³; $d_{crit} = 1.22 \times 10^{-2}$; $w_{crit} = -0.82$ cm/s .

2.5.2 Oseen's relationship

On the basis of theoretical investigations, Oseen (in Németh 1963) extended the validity ranges of the Stokes equation by applying a theoretically derived correction. The essence of this correction is connected to the partial consideration of the convective acceleration term of the Navier–Stokes equation when solving the differential equation. The drag force obtained with Oseen's correction is expressed as

$$W = 6\pi r\eta w \left(1 + \frac{3}{8} \frac{wr}{v} \right) = 6\pi r\eta w \left(1 + \frac{3}{16} \frac{wd}{v} \right) = 6\pi r\eta w \left(1 + \frac{3}{16} \mathrm{Re} \right). \tag{2.27}$$

Equating (2.16) and (2.27), the new expression for resistance coefficient C_w is obtained as

$$C_w = \frac{24}{\mathrm{Re}} \left(1 + \frac{3}{16} \mathrm{Re} \right). \tag{2.28}$$

The upper limit of applicability of Oseen's relationship is approximately $\mathrm{Re} = 5$.

The curve constructed by using eq (2.28) fits well, in the $\mathrm{Re}\leq 5$ range, with the experimentally determined $C_w = f(\mathrm{Re})$ curve.

The settling velocity obtained on the basis of Oseen's correction is:

$$w = \frac{-\dfrac{6\eta g}{d} + \sqrt{\dfrac{36\eta^2 g^2}{d^2} + \dfrac{3}{2}\gamma(\gamma_1 - \gamma)gd}}{\dfrac{9}{4}\gamma} ; \qquad (2.29a)$$

$$w = \frac{8}{3}\left[\sqrt{\left(\frac{v}{d}\right)^2 + \frac{1}{24}\frac{\gamma_1 - \gamma}{\gamma}gd} - \frac{v}{d}\right]. \qquad (2.29b)$$

2.5.3 Goldstein's relationship

The correction proposed by Goldstein (in Németh 1963) further refined the calculation method for C_w and the resulting curve fits the experimental $C_w = f(\mathrm{Re})$ curve over a wider range. According to Goldstein, the corrected expression for the resistance coefficient is

$$C_w = \frac{24}{\mathrm{Re}}\left(1 + \frac{3}{16}\mathrm{Re} - \frac{19}{1280}\mathrm{Re}^2 + \frac{71}{20{,}480}\mathrm{Re}^3 - \ldots\right). \qquad (2.30)$$

$\mathrm{Re}\leq 8$ can be accepted as the range of validity of this equation. It is to be noted that calculations with the above equation are rather cumbersome, and thus it cannot be proposed for practical use.

2.5.4 Hazen's relationship

Hazen's formula (in Ivicsics 1968) defines a direct relationship between settling velocity w (cm/s), particle size d (cm) and the liquid temperature T (°F):

$$w = w_{T\,(\mathrm{F}^\circ)} = 100\, d\frac{10 + T}{60}. \qquad (2.31)$$

The range of validity is $0.1 < d$ (mm) < 1.0.

2.5.5 Allen's relationship

Allen (in Ivicsics 1968) proposed a semi-empirical formula for calculating the resistance coefficient:

$$C_w = \frac{18.5}{\mathrm{Re}^{0.6}}. \qquad (2.32)$$

This equation is widely utilized for practical purposes in the range of $1.0 < \mathrm{Re} < 500$.

He also proposed another relationship for the calculation of C_w:

$$C_w = \frac{5\pi}{4\sqrt{Re}}. \tag{2.33}$$

The experimentally proven range of validity is $30 \leq Re \leq 300$. Calculating with average C_w values in the above interval, the Allen formula can also be derived from Newton's equation:

$$w = 109\,d(\varrho_1 g - 1)^{2/3} \quad (cm/s) \tag{2.34}$$

eq (2.34) corresponds to a water temperature of $10\,°C$.

2.5.6 Goncharov's relationship

Goncharov (in Velikanov 1955) proposed three formulae for calculating the settling velocities — in three different velocity ranges — of particulate matter:

$$w = 40.6\,\frac{(\gamma_1 - \gamma)d^2}{\eta} \tag{2.35a}$$

if $0.001 < d\,(cm) < 0.015$,

$$w = 67.7\,\frac{(\gamma_1 - \gamma)d}{\gamma} + \frac{(\gamma_1 - \gamma)}{1.92\gamma}\left(\frac{T}{26} - 1\right) \tag{2.35b}$$

if $0.015 < d\,(cm) < 0.15$; and

$$w = 33.1\,\frac{(\gamma_1 - \gamma)d}{\gamma} \tag{2.35c}$$

if $d\,(cm) > 0.15$.

In eq (2.35b) the effect of water temperature T $(°C)$ is expressed directly.

2.5.7 Simple semi-empirical relationships

Results of investigations carried out by various researchers have proved that in the turbulent range of settling the velocity can be well estimated by the expression

$$w = \text{const}\,d^a, \tag{2.36}$$

where const and a are constants to be determined for a given case; const includes the effects of all factors except the characteristic particle size, that affect settling velocity. Some concrete examples will be presented below.

For *spherical particles* const$=22.7$ and $a=0.5$. If particle diameter d is substituted in mm units in the equation, then the settling velocity is obtained in cm/s dimension. According to the measurements by Sokolov (in Velikanov 1955), const$=36$ and $a=0.5$ for a water temperature $T=10\,°C$, and for the Reynolds number range $30 < Re < 800$. It is to be noted that the settling velocity of particles of shapes other than spherical can also be calculated by the above expression if d is substituted for by the nominal or mean particle diameter d_n.

Fair et al. in 1968 investigated the settling velocity of particles in sewage water. For a water temperature of 10 °C the following expressions were obtained:

$$w = 14d \ (\text{cm/s}), \quad \text{if} \quad d > 0.22 \ (\text{mm}) \tag{2.36a}$$

$$w = 67d^2 \ (\text{cm/s}), \quad \text{if} \quad d < 0.22 \ (\text{mm}). \tag{2.36b}$$

Equations of the type of (2.36) are frequently used in the field of settling and sedimentation. Their main *advantage* is their simplicity and easy handling. Their *disadvantage* is that the range of validity is generally narrow and refers to experimental conditions only.

2.5.8 Rubey's relationship

Considerable efforts have been made to construct relationships that could be used over very wide ranges of the settling velocity. In this context the relationship by Rubey (in Ivicsics 1968), also proved experimentally, can be mentioned.

Rubey's settling formula is:

$$w = \sqrt{\frac{2}{3}\varrho^* gd + \frac{36\eta^2}{\varrho^2 d^2}} - \frac{6\eta}{\varrho d}, \tag{2.37}$$

where $\varrho^* = \dfrac{\varrho_1 - \varrho}{\varrho}$.

At low d and high η values, i.e., at low Re numbers, eq (2.37) takes the Stokes equation form. For high Re numbers

$$w = \sqrt{\frac{2}{3}\varrho^* gd}. \tag{2.38}$$

Rubey's relationship can also be written in dimensionless form. After appropriate rearrangement, one obtains that

$$\text{Fr Re} + 12 \text{Fr} = \frac{2}{3}\varrho^* \text{Re}, \tag{2.39}$$

where $\text{Fr} = \dfrac{w^2}{gd}$ and $\text{Re} = \dfrac{wd}{v}$.

For practical calculations, a validity range of $0.002 \leq d\,(\text{cm}) \leq 10$ can be considered.

2.5.9 Description of the full range of settling velocity

Efforts made towards the mathematical formulation or graphical presentation of processes over the full range of settling velocities and particle dimensions were obvious.

Among the expressions for the full range of the processes, the physically-based general relationship of Newton (eq 2.18) and McGauhey's equation (in Ivicsics 1968), derived from the general relationship $F(W, w, d, \varrho, \eta)=0$, with the methods of dimension analysis, can be mentioned:

$$w^n = \frac{4}{3}\frac{g}{k}\left(\frac{\varrho_1-\varrho}{\varrho^{n-1}}\right)\eta^{n-2}d^{3-n} \tag{2.40}$$

and

$$C_w = k\,\mathrm{Re}^{n-2}, \tag{2.41}$$

where n and k are constants over certain ranges of the settling process. For example, with $n=1$ and $k=24$, Stokes' law is obtained. Further, by substituting $n=2$ and $k=C_w$, Newton's law can be derived. In the transitional range, the condition of $1<n<2$ prevails.

The widening of the limits of the calculations was the objective of the already mentioned corrections by Oseen and Goldstein, and Rubey's efforts were also focussed on this. Fair's equation, frequently referred to in the American literature, is also based on similar principles:

$$C_w = \frac{24}{\mathrm{Re}} + \frac{3}{\sqrt{\mathrm{Re}}} + 0.34. \tag{2.42}$$

Equation (2.42) is valid for spherical particles and in the range of $0.5<\mathrm{Re}<10^4$.

For practical calculations mainly graphical or grapho-analytical procedures can be proposed. Based on the considerations in the previous section, it becomes evident that in the Stokes range w is proportional to d^2, but in the Newton range to $d^{1/2}$. According to this, the relationship $w=f(d)$ can be plotted as shown in *Fig. 2.3*, which can be considered the characteristic diagram of the settling process. In this case, the transitional phase is represented by Allen's relationship. It is to be noted that in constructing the lower part of the $w=f(d)$ curve, the effects of Brownian motion were also taken into consideration.

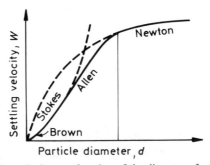

Fig. 2.3. Settling velocity as a function of the diameter of spherical particles

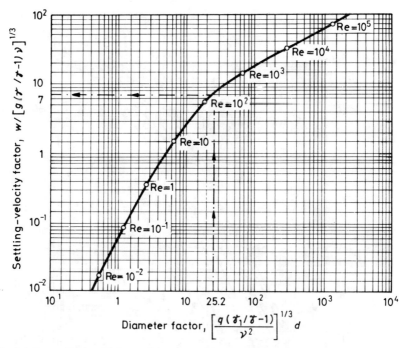

Fig. 2.4. Determination of the settling velocity of spherical particles in stagnant water,
as a function of the diameter coefficient
(after Fair, Geyer and Okun 1968)

In *Fig. 2.4* a graphical nomogram is presented for the estimation of diameter d or settling velocity w of an isometric particle, in the range of $10^{-2} < \mathrm{Re} < 10^{5}$, as derived grapho-analytically.

The theoretical principle of the diagram is based on eq (2.18b), which can be rearranged to obtain

$$\left(\frac{3}{4} \mathrm{Re}^2 C_w\right)^{1/3} = \left[\frac{g(\varrho_r - 1)}{v^2}\right]^{1/3} d \qquad (2.43)$$

and

$$\left(\frac{4}{3}\frac{\mathrm{Re}}{C_w}\right)^{1/3} = \frac{w}{[g(\varrho_r - 1)v]^{1/3}}, \qquad (2.44)$$

where

$$\varrho_r = \frac{\varrho_1}{\varrho} = \frac{\gamma_1}{\gamma} = \gamma_r$$

is the relative density and specific weight.

It is to be noted that the above relationship is similar to eqs (2.20) and (2.21).
The right-hand-side terms of eqs (2.13) and (2.44) can be considered to be the

3

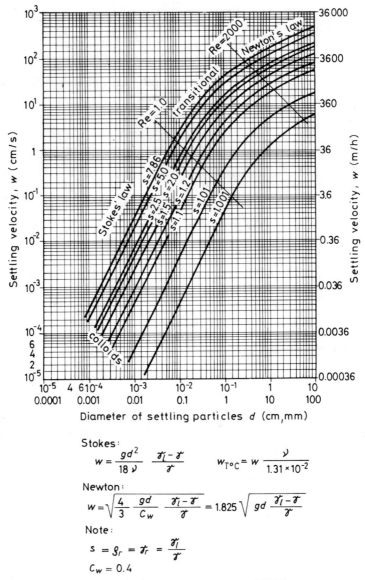

Stokes:
$$w = \frac{gd^2}{18\,\nu}\ \frac{\gamma_i - \gamma}{\gamma} \qquad w_{T°C} = w\ \frac{\nu}{1.31 \times 10^{-2}}$$

Newton:
$$w = \sqrt{\frac{4}{3}\ \frac{gd}{C_w}\ \frac{\gamma_i - \gamma}{\gamma}} = 1.825\ \sqrt{gd\ \frac{\gamma_i - \gamma}{\gamma}}$$

Note:
$$s = g_r = \gamma_r = \frac{\gamma_i}{\gamma}$$
$$C_w = 0.4$$

Fig. 2.5. Settling velocity of spherical particles in stagnant water of 10 °C temperature as a function of the particle size (after Fair, Geyer and Okun 1968)

"diameter factor" and the "velocity factor", respectively. The diagram of *Fig. 2.4* shows the relationship between these two factors on the basis of the above expressions. The steps in using this diagram in practice are as follows.

If one wants to estimate the settling velocity, then first the diameter factor is calculated on the basis of data on the respective variables. Next, the velocity factor is determined from the graph as a function of the diameter factor. This latter allows the direct calculation of settling velocity w. As an alternative, particle size d can also be estimated in a similar way. It is to be noted that the approximate value of the Re number can be also obtained from the graph, making use of the given intermediate ranges, thus enabling the estimation of C_w by utilizing the left-hand-side term of eqs (2.43) or (2.44).

In *Fig. 2.5* a series of curves is presented for the direct calculation of the settling velocity of spherical particles in stagnant water of 10 °C temperature. The parameter of the curves is relative density ϱ_r. When calculating for temperatures other than 10 °C, multiplication by $v(1.31 \times 10^{-2})$ is needed, where v is the kinematic viscosity corresponding to the desired temperature T (°C). The graph also indicates the characteristic ranges of settling.

Finally, in *Fig. 2.6* curves (adopted from various researchers) for the description of the settling velocity of spherical particles are presented. In the range of $d > 0.5$ the differences are quite significant. Nevertheless, the figure offers a good basis for comparing the results of different researchers.

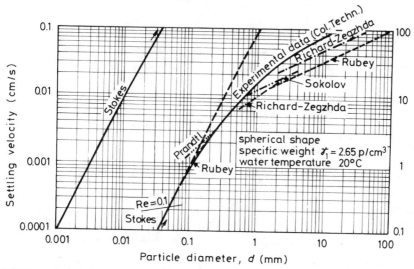

Fig. 2.6. Settling velocity of spherical particles as a function of the particle diameter; comparison of various research results (after Bogárdi 1972)

2.5.10 Summary of the characteristic ranges
of settling

According to the type of flow and from the dynamic point of view, three major ranges can be distinguished (see *Fig. 2.7*):

a) The *laminar phase*: Re<0.6 (Re<0.5 or Re<1.0).
In respect to the flow pattern, streamlines are symmetrical in the space before and after the settling body; the flow is laminar. Friction forces play the dominating role in forming the drag force.

b) The *transitional phase*: 0.6<Re<50.
There is a symmetrical pattern of vortices forming behind the settling particle. The thickness of the laminar sheet decreases with the increase of the Re number. Friction and inertia play equally important roles in forming the drag force.

c) The *turbulent phase*: Re>50.
Vortices of irregular pattern are formed behind the particles. The thickness of the laminar sheet continues to decrease. Inertia forces become dominant.

On the basis of the shape of the $C_w = f(Re)$ curve and that of the *calculation* of C_w, the following main ranges can be defined:

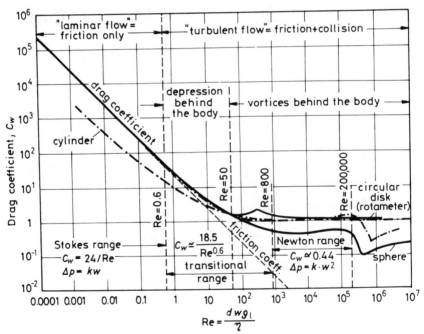

Fig. 2.7. Characteristic ranges of the settling process with the respective $C_w = f(Re)$ relationships (after Tettamanti 1970)

a) The *Stokes range*: $Re < 0.6$ ($Re < 0.5$ or $Re < 1.0$).
C_w is calculated by eq (2.25).

b) The *transitional range*: $0.6 < Re < 800$ ($Re < 2000$).
C_w can be calculated by eq (2.32) or by eq (2.42).

c) The *Newtonian range*: $800 < Re < 2 \times 10^5$.
In this case $C_w = 0.44$, or by rounding it off upward $C_w = 0.5$.

d) Abrupt local *minima* of the $C_w = f(Re)$ curve: $2 \times 10^5 < Re < 6 \times 10^5$.

The hydraulic explanation of this deviant range is that in the boundary layer around the body laminar flow changes suddenly over to turbulent. The range $Re > 2 \times 10^5$ has its significance mostly in the field of ballistics.

It is to be noted that Re ranges shown within brackets refer to cases where different authors consider different ranges of validity.

2.6 Settling of particles of shapes other than spherical

Practical problems are seldom related to the settling of spherical particles. In reality, the particles usually have other shapes. Nevertheless, the detailed description in Section 2.5 can be considered justifiable, as the mathematical, empirical and semi-empirical relationships derived for spherical particles provide a good basis for further investigations related to the introduction of *shape coefficients*.

2.6.1 Shape coefficients of settling particles

Shape coefficients defined in various ways are known. McNown and Malaika (in Ivicsics 1968) have, among others, presented a summary of these and described their role. These authors themselves improved the concept of shape coefficients in order to assure a more accurate and relatively simple characterization of particles.

Among the shape coefficients the following are worth mentioning: *sphericity*, *circularity*, *roundishness* and the *volumetric coefficient of Heywood*. Of these sphericity, which already has a classical role, will be discussed in more detail. In respect to the rest of the shape coefficients, the relevant literature should be referred to.

Definition of sphericity:

$$\zeta = \frac{\text{surface area of a sphere of a volume equal to that of the particle}}{\text{the actual surface area of the particle}}. \qquad (2.45a)$$

The shape coefficient can be calculated with the following expression:

$$\zeta = \frac{d_p}{d_s n} = \frac{d_p}{d_s \dfrac{f_p}{f_s}}, \qquad (2.45b)$$

where d_p is the mean diameter of the particle and d_s is the diameter of the sphere with a volume equalling that of the particle. Further

$$f_p = \frac{\text{the actual surface area of the particle}}{\text{the mass of the particle}} = \frac{\dfrac{d_s^2 \pi}{\zeta}}{\dfrac{d_s^3 \pi}{6} \varrho} \qquad (2.46)$$

and

$$f_s = \frac{\text{surface area of a sphere of identical volume}}{\text{mass of a sphere of identical volume}} = \frac{d_p^2 \pi}{\dfrac{d_p^3 \pi}{6} \varrho}. \qquad (2.47)$$

Consequently,

$$n = \frac{f_p}{f_s} = \frac{\text{specific surface area of a sphere of identical volume}}{\text{specific surface area of the particle}}. \qquad (2.48)$$

The sphericity of a spherical particle is, according to eq (2.45a), unity. For particles of shapes other than spherical, $\zeta < 1$. For example, the sphericity of a cube is $\zeta = 0.806$. It is to be noted that the mean particle diameter d_p should be determined experimentally (e.g., by sieve analysis).

Equivalent diameter d_e is also frequently used in practice. In one approach, d_e is the diameter of a sphere having a settling velocity identical to that of the particle. In another approach, d_e is the diameter of a sphere of volume $V = M/\varrho_1$, where M is the mass of the particle

$$d_e = \sqrt[3]{\frac{6}{\pi} V} = 1.24 \sqrt[3]{V} = 1.24 \sqrt[3]{\frac{M}{\varrho_1}}. \qquad (2.49)$$

Knowing the shape coefficient, the $C_w = f(\text{Re}, \zeta)$ relationship can be presented graphically, for particles of differing shapes, similarly to *Fig. 2.1*. In this approach, the shape coefficient becomes one of the parameters of the graph. *Figure 2.8* is presented as an example of this, where sphericity is the shape coefficient considered.

In the validity range of Stokes' law, the ratio of resistance coefficient C_w^* of a particle of a shape other than spherical to the C_w of a sphere can be calculated as a function of sphericity ζ using the following empirical formula:

$$\frac{C_w^*}{C_w} = \frac{1}{0.843 \log(\zeta/0.065)}, \qquad (2.50)$$

where C_w is that given by eq (2.25). In the turbulent phase of settling a simple linear relationship can be used:

$$C_w^* = 5.31 - 4.88\zeta. \qquad (2.51)$$

For spherical particles $\zeta = 1$ and thus $C_w^* = C_w = 0.43 \ (\approx 0.44)$.

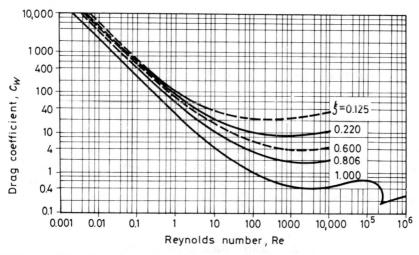

Fig. 2.8. Drag coefficient for particles other than spherical ones as a function of the Reynolds number and the coefficient of sphericity (after Ciborowski 1969)

2.6.2 Calculation of settling velocity on the basis of the relationship $C_w = f(Re; \zeta)$

If the relationship $C_w = f(Re; \zeta)$ is known, the settling velocity of particles of irregular shape can be well estimated in the validity range of Newton's squared resistance law, by the following expression:

$$C_w A \varrho \frac{w^2}{2} = C_w^* A^* \varrho \frac{w^{*2}}{2}$$

whence

$$w^* = w \sqrt{\frac{A C_w}{A^* C_w^*}} = w \sqrt{\zeta \frac{C_w}{C_w^*}} . \tag{2.52}$$

(Mark * refers to particles of irregular shape.)

2.6.3 The modified Stokes relationship

In the laminar phase of settling, the settling velocity of particles of irregular shape can be calculated on the basis of Stokes' relationship (eq 2.24) and the correction related to sphericity (eq 2.50), by the following expression:

$$w = \frac{gd^2}{18\eta} (\varrho_1 - \varrho) \left[0.843 \log \left(\frac{\zeta}{0.065} \right) \right] ; \tag{2.53}$$

eq (2.53) can be considered a modified Stokes relationship.

2.6.4 The Newton–Rittinger relationship

On the basis of the experimental $C_w = f(Re)$ curve for spherical settling particles, it is found that $C_w \approx 0.5$ in the range of $600 \leq Re \leq 2 \times 10^5$. With this assumption, the general eq (2.18b) simplifies into

$$w = \text{const}_1 \sqrt{d \frac{\varrho_1 - \varrho}{\varrho}} = \text{const}_2 \sqrt{d(\gamma_1 - 1)} \quad \text{(cm/s)}. \tag{2.54}$$

This expression is widely used for the calculation of settling velocities of particles of irregular shapes. In this case, mean diameter d_m should be substituted. This can be obtained experimentally, by sieve analysis. The experimentally derived value of const_2 is 55 for spherical particles, 35.7 for round shaped particles, 32.6 for cubic particles, 26.0 for elongated particles and 25.0 for flat particles.

2.6.5 Zegzhda's relationship

Velikonov and Zegzhda (1949) investigated the settling velocities of spherical and near-spherical particles over wide ranges of the settling process. Making use also of literature data, they derived a relatively general and thus quite complex empirical relationship:

$$\lambda = 0.105 + 4.5 \, Re^{-1} + \frac{2 \cdot 2.655}{\pi} \arctan Re^{-1/2}, \tag{2.55}$$

where

$$\lambda = \frac{gr\left(\dfrac{\varrho_1 - \varrho}{\varrho}\right)}{w^2} \quad \text{and} \quad Re = \frac{rw}{v}.$$

Equation (2.55) and its graphical representation (*Fig. 2.9*) substantially enhance the calculation.

In a special case, for low Reynolds numbers

$$\lambda = 2.76 + 4.5 \, Re^{-1}, \tag{2.55a}$$

and for high Re numbers

$$\lambda = 0.105. \tag{2.55b}$$

From eq (2.55) the value of w can be obtained by iterative approximation.

In using eq (2.55) for particles differing slightly from spherical, r should be substituted for by *equivalent radius* r_e, that can be obtained as

$$r_e = r, \qquad \text{if} \quad r < 0.125 \text{ mm} \quad \text{and}$$

$$r_e = \frac{2}{3} r^{2/3}, \quad \text{if} \quad r > 0.22 \text{ mm}.$$

Equivalent radius r_e is defined as the radius of a sphere having the same settling velocity as that of the particle concerned.

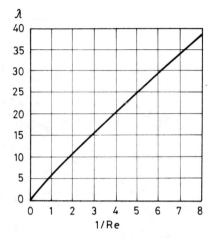

Fig. 2.9. Zegzhda's number as a function of Re^{-1} (after Velikanov 1949)

2.6.6 Description of the full range of settling velocities

Graphical aids have been elaborated to describe the settling process over the full range of settling velocities, also for particles of irregular shape. In relation to *Fig. 2.4*, it was already mentioned that this diagram can be used for both spherical and near-spherical, isometric bodies, in calculating the settling velocities. A further graphical aid is presented in *Fig. 2.10* where the respective curves are: No. 1 and No. 6 for spherical, No. 2 for rounded, No. 3 for angular, No. 4 for elongated and No. 5 for lamelliform particles.

An interesting feature of this nomogram is that the variables are dimensionless — according to the concept of Lyashchenko. The notations and the use of the nomogram can be formulated as follows.

The settling properties of individual particles that settle in stagnant liquidous media can be characterized by the following dimensionless numbers:

Reynolds number

$$Re = \frac{wd}{v};$$

Archimedes number

$$Ar = Ga\,\frac{\varrho_1 - \varrho}{\varrho} = \frac{Re^2}{Fr}\,\frac{\varrho_1 - \varrho}{\varrho} = \frac{gd^3(\varrho_1 - \varrho)\varrho}{\eta^2}; \qquad (2.56)$$

Lyashchenko number

$$Ly = \frac{Re^3}{Ar} = \frac{Fr\,Re\,\varrho}{\varrho_1 - \varrho} = \frac{w^3\varrho^2}{\eta(\varrho_1 - \varrho)g}, \qquad (2.57)$$

where Ga is the Galilei number.

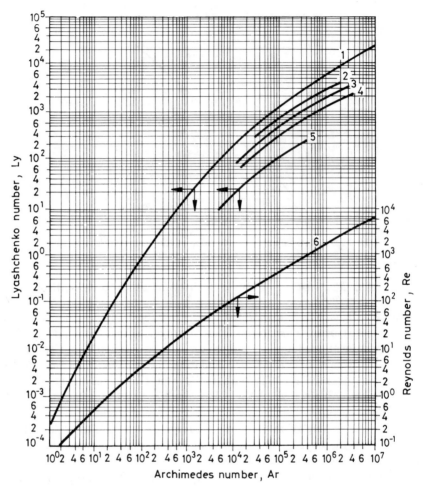

Fig. 2.10. Calculation of characteristic settling parameters, on the basis of dimensionless variables, for particles of different shapes, in stagnant medium (after Pavlov, Romankov and Noskov 1972)

The steps in calculating the settling velocity: The Ar number is calculated by eq (2.56); next Ly is obtained from *Fig. 2.10* as a function of Ar, eq (2.57) yield, after rearrangement, the Re value. Finally, the settling velocity is obtained as

$$w = \sqrt[3]{\frac{Ly\,\eta(\varrho_1 - \varrho)g}{\varrho^2}} = \frac{Re\,\eta}{d\varrho}. \qquad (2.58)$$

In an inverse case the dimension of a settling particle *d* can also be calculated. In this case, first Ly is calculated by eq (2.57) and then Ar is obtained from the nomogram. Finally, eq (2.56) yields, after rearrangement, diameter d. If particles of shapes other than spherical are concerned, then the $d=d_e$ substitution should be applied, using eq (2.49) presented above.

According to Russian literature, the range of validity of Stokes' settling law is characterized by the following conditions: $Re < 0.2$; $Ar < 3.6$; $Ly < 2 \times 10^{-3}$. In practice, however, settling is considered laminar up to the ranges of $Ar \approx 18$ and $Ly \approx 5 \times 10^{-2}$.

Finally, it should be noted that the physical basis of the dimensionless calculation is still related to the general relationship (2.18b). It also follows from this reasoning that the dimensionless numbers of eq (2.56) and eq (2.57) correspond to the eqs (2.20) and (2.21), respectively.

2.7 Obstructed settling

If particulate or floc-like particles come close together, *interactions* might occur. A similar effect is the so-called *wall-effect* that occurs at the walls of settling tanks and other structural elements.

In the course of the obstructed settling, several mechanisms are acting. Such effects are, for example: when the settling velocities of individual particles or flocs are changed upon their touching each other. Impact and friction forces are generated among the particles. Indirect interaction results from local movements of the liquid that result from the collision of the particles. This motion — on average — has a direction opposite to that of the settling. As a result of these processes, the velocity distribution and velocity gradients will change in the vicinity of the particles. Another indirect effect will be related to the changing of the physical properties of the flowing medium, such as viscosity and density. Finally, flocculation and coagulation processes, with their well-known substantial effects, will play a dominating role in forming the motion relations of individual particles.

These mechanisms are based on the action and reaction of various forces that act partly unidirectionally and partly opposite to each other, and vary in both space and time. Consequently, the exact calculation of obstructed settling velocities also faces theoretical problems.

In formulating the relationships discussed in the foregoing chapter, the motion characteristics of individual particles were considered. Consequently, their range of validity refers also to the range of free, unobstructed settling. Experience indicates, however, that by taking certain correction factors into account, the classical settling laws are well applicable, in certain cases, to the estimation of settling velocities under obstructed conditions, too. In this context, some calculation methods will be presented below. It is to be emphasized, however, that for solving practical problems in the range of obstructed settling, direct reliance on experimental results will yield more reliable results.

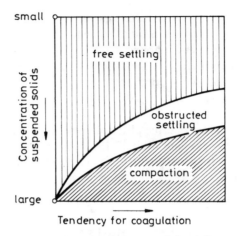

Fig. 2.11. Illustration of various zones of the settling process

Taking the above aspects into consideration, three different settling zones can be distinguished: a) free settling, b) obstructed settling, and c) compressive settling. In this latter zone, compression of the already settled substances (the sludge) also takes place. The relative position of the above three zones is illustrated by *Fig. 2.11.*

2.7.1 The modified Stokes relationship

It was experimentally proved that Stokes' law on settling, as described in Section 2.5.1, remains well applicable also in the obstructed zone of settling, for suspensions consisting of small particles. In this case, however, the physical characteristics of the suspension should replace density and viscosity in the equation. It is to be noted that for fine suspensions, dynamic viscosity η_s can be determined by Einstein's relationship:

$$\eta_s = \eta\left(1 + 2.5\,\frac{x}{\gamma_1}\right) = \eta[1 + 2.5(1 - \varepsilon)], \qquad (2.59)$$

where

$$\varepsilon = \frac{V_{\text{suspension}} - V_{\text{susp. matter}}}{V_{\text{suspension}}} = \text{the relative volume (``pore volume'') of the sus-}$$

pension,

η is the dynamic viscosity of the liquid when containing no suspended matter,

x is the suspended material content of the suspension, and

γ_1 is the specific weight of suspended matter.

Density ϱ_s of the suspension will also be required and this can be calculated by the following expression:

$$\varrho_s = \varrho_1(1 - \varepsilon) + \varrho\varepsilon. \qquad (2.60)$$

Knowing η_s and ϱ_s, the Stokes equation can be rewritten as

$$w_f = \frac{gd^2}{18\eta_s}(\varrho_1 - \varrho_s),\qquad(2.61\text{a})$$

where w_f is the settling velocity of particles in the suspension (relative velocity as compared to that of the liquid).

From eq (2.60), the substitution $(\varrho_1 - \varrho_s) = \varepsilon(\varrho_1 - \varrho)$ can be applied. Furthermore, the introduction of w^*, the relative obstructed settling velocity as related to the position of the wall of the settling facility, is desirable instead of w_f. As the settling flux of matter in a unit volume of the suspension can be expressed as $(1 - \varepsilon)w^*$, and the volumetric flux of displaced liquid flowing upward is $\varepsilon(w_f - w^*)$, the respective mass-balance equation yields $w^* = \varepsilon w_f$. Velocity difference $(w_f - w^*)$ is the flow velocity of the liquid, as related to the position of the wall.

With the above knowledge, eq (2.61) can be rewritten as

$$w^* = \frac{gd^2}{18\eta}(\varrho_1 - \varrho)\varepsilon^2 f(\varepsilon) = w\varepsilon^2 f(\varepsilon),\qquad(2.61\text{b})$$

where w is the settling velocity, to be derived from the original Stokes equation, and w^* is the obstructed settling velocity as modified on the basis of Stokes' law.

2.7.2 Steinour's relationship

According to the procedure published by Steinour in 1944 one may derive the obstructed settling velocity from Stokes' equation by introducing a correction function $F(\varepsilon)$. In this case, the ratio of w^*, for the obstructed phase of settling to w, corresponding to the original Stokes equation, equals the function $F(\varepsilon)$, that is;

$$\frac{w^*}{w} = F(\varepsilon) = \frac{1}{143}\varepsilon^2 \, e^{-4.2\,(1-\varepsilon)}.\qquad(2.62)$$

Steinour's relationship can also be written in the following form:

$$w^* = wF(\varepsilon) = w\varepsilon^2 f(\varepsilon),$$

where

$$f(\varepsilon) = 10^{-1.82\,(1-\varepsilon)} \approx 0.123\,\frac{\varepsilon}{1-\varepsilon}.\qquad(2.63)$$

Equation (2.63) can be applied in the range of $\varepsilon < 0.7$.

2.7.3 Calculation on the basis of the fluidized-bed theory

The laws of obstructed settling can also be discussed on the basis of the well-known relationships of the fluidized-bed theory. One of the approaches might be the application of the Darcy or Koženy–Carman relationship to laminar-flow conditions, when the set of settling particles is considered as a porous medium and the relationship between "seepage velocity" and the hydraulic head can be thus formulated. The use of this method is described in detail in the book by Fair, Geyer and Okun (1968).

Based on Darcys' law the relationship between seepage velocity v_s and the effective mean velocity v_{eff}, within the pores, can be formulated as $v_s = n v_{\text{eff}}$. In the case of obstructed settling, this relationship refers to $w^* = \varepsilon w_f$, as presented above. Further, it was experimentally proved that the following expression gives a good approximation:

$$\frac{w^*}{w} = \varepsilon^4 \tag{2.63a}$$

2.7.4 Further relationships for obstructed settling

In this section, some further nomograms for the calculation of obstructed settling velocities will be presented. *Figure 2.12* demonstrates, after Camp (1953), the decrease of settling velocity, during obstructed settling in stagnant water, as a function of the volumetric concentration and the Re number, on the basis of data obtained from various authors. It can be seen that velocity ratio w^*/w also decreases with increase of the Reynolds number. Furthermore, the volumetric concentration should exceed 0.5% in order to show a considerable decrease in the obstructed settling velocity. The above threshold value means concentrations of suspended solids of about 1000— 5000 mg/l in river water, while for municipal sewage water, in lime softeners, the respective range of $CaCO_3$ is 1000—2000 mg/l.

From the nomogram in *Fig. 2.13*, frequently used in chemical-engineering practice, velocity ratio

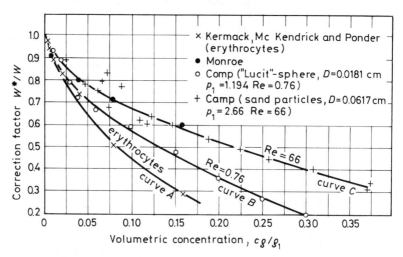

Fig. 2.12. Decrease of settling velocity at obstacled settling (after Camp 1953)

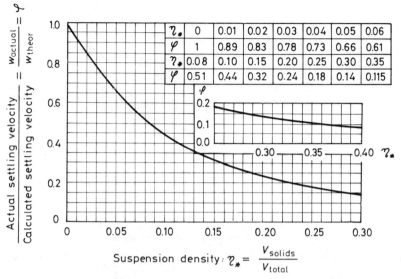

Fig. 2.13. Effect of suspension density on settling velocity (after Tettamanti 1970)

$$\varphi = w_{\text{actual}}/w_{\text{theoretical}} = w^*/w$$

can be obtained as a function of the volumetric density (concentration) of the suspension. This actually yields the obstructed settling velocity.

Two further figures will be presented, after J. Bogárdi, also making use of the measurement data of McNown and Lin and Schoklitsch (in Bogárdi 1972). *Figure 2.14* shows the ratio of settling velocities in clean water and in water containing suspended solids, as a function of the weight percentage of suspended solids, for quartz sand. The structure of this figure is identical with that of Camp (1942b), with the exception that in the vertical axis w/w^* is plotted in the former. The figure indicates that w/w^* decreases with increasing Re number, that is w^*/w increases also in this case. Finally, *Fig. 2.15* illustrates the relationship between settling velocity and the particle size, for suspensions of varying concentrations C and specific weight γ_s, at water temperatures of about 10 °C. Specific weight is defined by the following formula:

$$\gamma_s = \gamma - \frac{\gamma}{\gamma_1} C + C = \gamma + C \left(1 - \frac{\gamma}{\gamma_1} \right). \qquad (2.64)$$

2.7.5 Mean settling velocity of inhomogeneous particles

In practical cases, the tasks are usually related to the settling of materials having an inhomogenous particle composition. In this case it seems to be desirable to determine the mean settling velocities w_i for each particle fraction. Denoting the

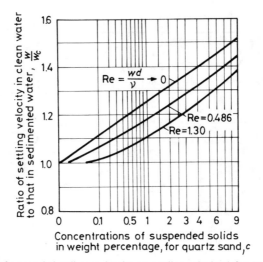

Fig. 2.14. Effect of suspended sediment density on settling velocity (after McNown and Lin, in Bogárdi 1972)

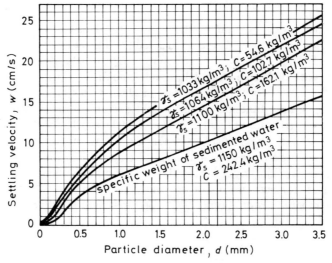

Fig. 2.15. Settling velocity of quartz particles as a function of the particle diameter and the concentration of solids (after Schoklitsch, in Bogárdi 1972)

percentage weight of the i-th fraction by p_i, the approximate value of the mean settling velocity, for the total system, can be obtained as

$$w_k = \frac{\Sigma(w_i p_i)}{100}.$$ (2.65)

The velocity of particles of differing diameters (d_1, d_2) can also be identical — a case termed as joint settling $(w_1 = w_2)$. The condition at which joint settling occurs in the Stokes range of settling is expressed as

$$\left(\frac{d_1}{d_2}\right)^2 = \frac{\varrho_2 - \varrho}{\varrho_1 - \varrho},$$ (2.66)

while in Newton's turbulent range $(C_w = \text{const})$ this condition becomes

$$\frac{d_1}{d_2} = \frac{\varrho_2 - \varrho}{\varrho_1 - \varrho}.$$ (2.67)

2.8 Settling of flocculent substances

In water and sewage treatment, the task is frequently to solve the settling of flocculent suspended matter. Characteristic examples are: activated sludge, precipitates resulting from chemical treatment, settling of flocs, etc.

2.8.1 Settling of flocculent substances of low concentrations

Reviewing the expressions derived for the settling properties of particulate matter, it is found that, for a given particle, the settling velocity does not change with time (or with depth of settling). The case is different with flocculent substances, when the dimension of flocs changes (due to flocculation) in time and thus with the depth of the settling facility. Consequently, in the case of flocculent substances, the time factor (and thus the depth of settling) plays a dominating role, in addition to the floc size and to density relations. Smaller and larger flocs might become attached to each other, thus changing the floc size. For these reasons, the expressions derived for the settling velocity of flocculent substances are basically different from those of the particulate matter. In such cases, laboratory experiments are unconditionally required for obtaining accurate design parameters. In the subsequent paragraphs the execution of laboratory experiments will be described on the basis of the detailed experimental procedures of Eckenfelder and O'Connor (1961).

The experimental equipment, shown in *Fig. 2.16a*, consists of a settling cylinder, from which samples can be taken at different depths. The height of the cylinder should be about the same as the depth of the prototype settling facility, or rather it should be identical. At the start of the experiments the cylinder is to be filled with water or sludge. It is important to assure a homogeneous concentration and temperature

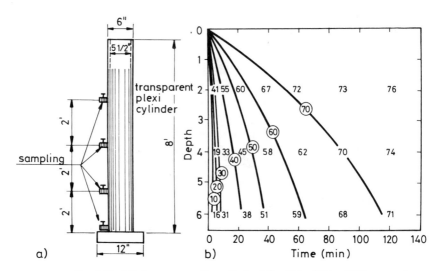

Fig. 2.16. a) Laboratory settling cylinder (after Eckenfelder 1970),
b) Settling efficiency curves

distribution, in order to avoid the occurrence of secondary currents. The efficiency of the settling process is defined as the difference between the initial and the actual concentrations in percentage of the initial concentration C_0 at time $t=0$. Evaluating the efficiencies obtained for different points of time and different depths, a series of curves, as shown in *Fig. 2.16b*, is obtained (for particulate matter this would display a series of straight lines). The tangent of these curves at a particular point yields the settling velocity corresponding to that point. It can be seen that the lower portions of the curves become steeper and steeper, indicating that the larger flocs in the lower layers move with higher settling velocities. The intersection of a curve with the horizontal axis marks the retention time that will be required to attain the settling efficiency represented by that curve, at the settling depth defined by the height of the cylinder.

2.8.2 Settling of flocculent substances of high concentrations

In the case of the settling of flocculent substances of high concentrations, the settling velocity changes significantly not only as a function of the changing floc size, but also of that of the interaction between the particles. Namely; at higher concentrations, suspended particles become compacted and attached to each other (obstructed settling and compaction).

For a given waste water, experimental methods should be used for determining the settling properties; consider the different cases shown in *Fig. 2.17*. With settling in a glass cylinder, different conditions correspond to different points of time. At the start

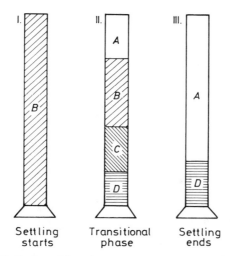

Fig. 2.17. Settling of flocculate substances of high concentrations

of settling (phase I), the homogeneous suspended solids concentration C_0 can be measured in the well-mixed medium. At the end of the settling process (phase III), the settled and compacted sludge will be clearly separated from the overlying water. Between these two extremes generally four different levels of differing properties can be distinguished (phase II):

a) The upper layer does not contain settleable solids and its depth is increasing with time;

b) The depth of the second layer is gradually decreasing. The sludge particles of this layer are settling with uniform velocity.

c) The third layer is a phase transitional between the second and the fourth layers. Here settling velocities are already decreasing due to the interaction of suspended particles. The depth of this layer gradually decreases with time.

d) The fourth layer contains the settled substance. The depth of the third and fourth layers, together, increases until the third layer becomes the upper layer containing no suspended particles. According to the theoretical considerations of Kynch (1952) this process occurs at a uniform rate.

The process can be better illustrated by plotting the state of the suspension as a function of time. Let us consider the conditions of phase II. The quantitative relations are demonstrated by *Fig. 2.18*. At a point in time t after the start of settling, four different zones, as already mentioned, can be distinguished in the glass cylinder. In the uppermost zone, the water (the waste water) contains practically no solids, and is clearly separated from zone B. The continuous curve of *Fig. 2.18* shows the time variation of this boundary line (the sludge level). Zone B contains suspended solids (sludge) that settle with uniform velocity. Zone C, that has a rising upper level as long

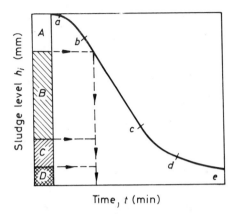

Fig. 2.18. Scheme for analyzing the settling process and compaction of flocculete substances

as zone *B* is present, is transitional to compression zone *D*. Function $h = f(t)$ can be also divided into four sections. In section *a—b* coagulation takes place and the flocs become attached to each other thus increasing the settling velocity along this section. It is to be noted that section *a—b* frequently prevails for only a very short time, for a few minutes. Section *b—c*, represented by a straight line in the figure, is characterized by a uniform settling velocity. During the period corresponding to section *c—d*, decreasing settling velocities can be observed, due to the increase of concentration. Finally, in the compression range of section *d—e*, the downward moving velocity of boundary line *A—D* gradually converges to zero.

2.9 Settling in a flowing medium

2.9.1 Effects of laminar and turbulent flow

The effects of laminar flow on the settling properties will be determined by the direction of flow. In a fluid flowing horizontally, the velocity of settling of solid (non-flocculate) particles will, theoretically, not change, due to the independent character of the acting forces. In a case when the flowing medium of mean velocity *v* has a vertical velocity component, the friction forces and shape factors of drag resistance will excercise their effects, and thus a correction must be applied.

In turbulent flow, the pulsation of velocities and pressures will also play a role in such a way that the settling velocity of certain particles increases, while that of other particles decreases. Statistical considerations lead to the above conclusions. Settling pathways also show statistical variations, accordingly. This is demonstrated by *Fig. 2.19*, showing the distribution of the distance of travel of settling particles that started from a given, identical point. It is to be emphasized that in calculating settling

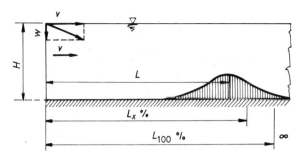

Fig. 2.19. Distribution of particles settled on a channel bed, when released from an identical point

properties for practical purposes, the decisive effects will be related to vertical velocity pulsation that reduces settling.

The effects of laminar and turbulent flow on the design of settling facilities will be discussed in more detail in Chapter 4. Only some experimental results — related to the effect of horizontal flow on the velocity of settling — will be presented in this chapter.

The method of Levin (in Ivicsics 1968) is based on direct consideration of a decrease of settling velocity. The relationship proposed by Levin is the following:

$$w_t = w - \sigma_t v ,\qquad (2.68)$$

where

$$\sigma_t = \frac{0.132}{\sqrt{m}} ,\qquad (2.69)$$

v is the horizontal flow velocity (mm/s), w_t is the settling velocity in turbulent flow (mm/s), and m is the depth of water in the settling tank (m). The use of this method can be mainly recommended in the case of higher flow and settling velocities (e.g. in sand traps).

On the basis of the measurement results of Savaljev (in Ivicsics 1968) a relationship similar to eq (2.69) can be obtained:

$$\sigma_t = \frac{0.0282}{\sqrt[5]{m}} .\qquad (2.70)$$

In Fig. 2.20 the relationship $w_t = f(v)$, derived on the basis of the measurement data of Ivicsics (1968) is shown. The linear relationship thus obtained is in harmony with eq (2.68).

Several researchers have suggested the use of various empirical equations in which the effect of turbulent flow is taken into account as correction. Such expressions are the following:

$$\alpha_t = \frac{w}{w - v_t} ;\qquad \alpha_t = \frac{aw}{w - b} \quad \text{and} \quad a_t = \frac{w}{w - cv} ,\qquad (2.71a\text{—}c)$$

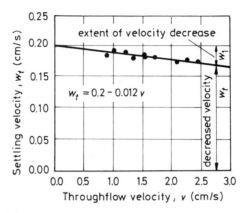

Fig. 2.20. Relationship between settling velocity and throughflow velocity (after Ivicsics 1968)

where

   v_t is the vertical velocity component of turbulent flow, and a, b, and c are empirical constants.

The value of α_t varies, generally, between 1.0 and 2.0. If turbulence causes favourable effects (i.e., it enhances flocculation), α_t might be smaller than 1.0. Accordingly, for $\alpha_t > 1.0$ the effect of turbulence is unfavourable from the viewpoint of settling.

2.9.2 Role of the velocity gradient

   The critical size of settling particles can also be defined as a function of the vertical velocity gradient dv/dz. Flocs start to disintegrate into smaller parts when the critical size is exceeded. Velocity gradients play an important role in the settling process of flocculate substances. This, subject will, however, be discussed in more detail in a separate section dealing with flocculation processes, as velocity gradients affect the settling process mostly through flocculation. Here, only the respective research work by Camp (1953) is referred to.

2.9.3 The effects of mixing

   According to experimental observations, a certain degree of mixing affects settling velocities favourably. This effect can best be explained on the basis of the flocculation theory. The effects of mixing on the settling process can also be proven experimentally. Moreover, relationships directly applicable in technological practice were mostly derived on the basis of experimental methods.

   Figure 2.21/a–b is presented as a good example (after Eckenfelder 1970) of the effects of mixing (among others). The most widely applicable parameter characterizing the settling process of flocculent substances of high concentrations is, as was

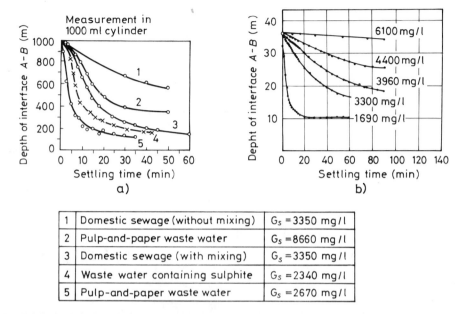

Fig. 2.21/a–b. Variation of the depth of interface as a function of the time and the effect of mixing, for different waste waters

1	Domestic sewage (without mixing)	$G_s = 3350$ mg/l
2	Pulp-and-paper waste water	$G_s = 8660$ mg/l
3	Domestic sewage (with mixing)	$G_s = 3350$ mg/l
4	Waste water containing sulphite	$G_s = 2340$ mg/l
5	Pulp-and-paper waste water	$G_s = 2670$ mg/l

mentioned above, the downward-moving velocity of the upper level of sludge, that is the boundary surface between the first and the second layers. It is clearly seen that stirring of appropriate intensity plays an important role in increasing the settling velocity. Namely, slow stirring acts favourably on the formation of flocs, thus enhancing flocculation (compare curves 1 and 3 in *Fig. 2.21/a* and *b*, where the initial solid concentrations G_s were identical). The initial concentration itself plays a decisive

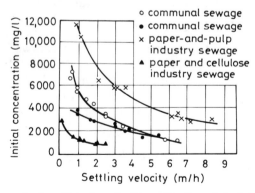

Fig. 2.21/c. The lowering velocity of interface A—B as a function of the initial concentration (after Eckenfelder and Melbinger 1957)

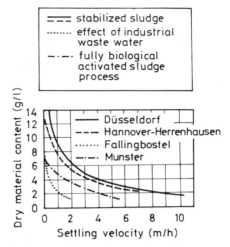

Fig. 2.21/d. Settling velocity of activated sludge originating from different municipalities as a function of the dry material content (after Stobbe 1964)

role, as with an increase of the initial concentration, the settling velocity decreases (due to the above-mentioned reasons) if all other experimental conditions are the same (compare curves 2 and 5). The results of theoretical and experimental investigations equally indicate that the settling velocity of flocculate substances of high concentrations depends (assuming identical experimental conditions) primarily on the initial concentration C_0. Obviously, experimental conditions (flow conditions, stirring, temperature, etc.) also affect the ways and means of the settling process.

Further information on settling velocities can be obtained from *Fig. 2.21/c* and *Fig. 2.21/d.*

2.9.4 Settling and fluidization

Phenomena related to fluidization occur in several water and waste-water treatment facilities, mainly in settling tanks (during the backwashing phase). In this respect, *Fig. 2.22*, which is similar to *Fig. 2.10*, is presented. This series of curves permits, for example, the calculation of the velocity required to establish a fluidized bed of pore volume ε, as a function of the diameter of particles, or vice versa. The validity range is $\varepsilon = 0.4$—1.0. The pore volume of the fluidized bed is, by definition

$$\varepsilon = \frac{V_f - V_r}{V_f},$$
(2.72)

where V_f is the volume of the fluidized layer and V_r is the total volume of solid particles. The two extremes of the graph — as related to the fluid layer — are the stagnant layer and the transport.

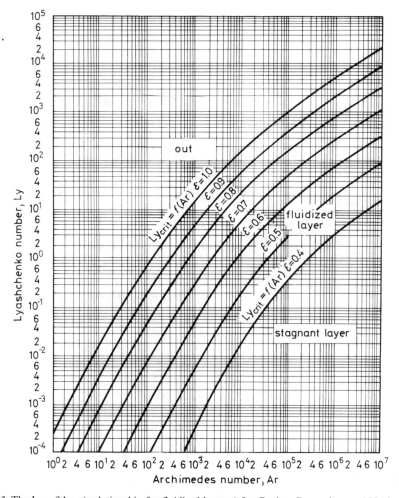

Fig. 2.22. The Ly = f(Ar; ε) relationship for fluidized layers (after Pavlov, Romankov and Noskov 1972)

2.10 The effect of temperature

In the above-described relationships, the effect of temperature can, mostly, be implicitly considered through the dependence of viscosity on temperature. The nomogram of *Fig. 2.5* is a good example of this. The temperature of the fluid is directly expressed in Hazen's formula (eq 2.31) and in Goncharov's (eq 2.35b) relationship.

It is to be noted that the change of density as a function of temperature is negligible. The temperature dependency of viscosity can be considered by the follow-

I. Mechanical treatment

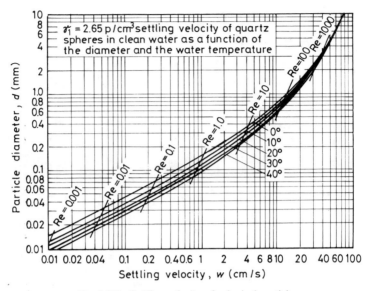

Fig. 2.23/a. Settling velocity of spherical particles

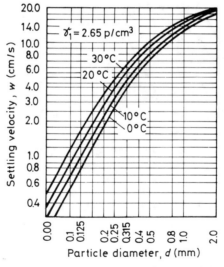

Fig. 2.23/b. Settling velocity of spherical sand particles as a function of the particle size and the water temperature

ing expression (the reference settling velocity being that corresponding to 20 °C):

$$w_r(°C) = \frac{w_{20}}{\frac{\eta_r}{\eta_{20}}}, \qquad (2.73a)$$

as velocity w is inversely proportional to η at a certain power (or to η in the Stokes range).

On the basis of eq (2.31), the following dependency on temperature $T(°C)$ can be derived:

$$w = w_T(°C) = 100\, d\, \frac{42 + 9/5\,T}{60}. \qquad (2.73b)$$

For example if $T = 10\,°C$, then

$$w = w_{10} = 100\, d. \qquad (2.74)$$

Consequently, the settling velocity corresponding to a desired temperature $T(°C)$ can be approximated as

$$w_r(°C) = w_{10}\, \frac{42 + 9/5\,T}{60} = w_{10}\, \frac{21 + 0.9\,T}{30}. \qquad (2.75)$$

According to other American authors, the following expression can be used for a wider range of practical applicability:

$$W_T(°C) = w_{10} \left(\frac{21 + 0.9\,T}{30} \right)^{1/2}. \qquad (2.76)$$

Finally, a series of curves is presented in *Fig. 2.23/a* in which the joint effect of particle diameter and temperature on the settling velocity of spherical quartz particles is specified in the Reynolds number range of $10^{-3} < \mathrm{Re} < 10^3$. *Figure 2.23/b* shows the similar results by Lane (in Horváth 1984).

2.11 Some additional data and diagrams

Figure 2.24 shows the $C_w = f(\mathrm{Re})$ relationships for bodies of various shapes immersed in water. In respect to part *b* of *Fig. 2.24* it is to be noted that the relationship is valid for elongate bodies as well. Resistance coefficients of prisms, cylinders, etc. can also be calculated. The respective practical examples are related to the investigation of the hydraulic effects of structural elements, beams, girders, etc. Analogous relationships are: $\lambda = f(\mathrm{Re})$ for pipes (the Nikuradse harp), the $C_w = f(\mathrm{Re})$ relationship for the settling process, and the $\zeta = f(\mathrm{Re})$ expression used for designing mixers.

The diagrams of *Fig. 2.25* present some further relationships useful in considering the effects of particle diameter, particle density ϱ_l, temperature and viscosity. According to Oehler (in Horváth 1984), this diagram can also be used for determining the settling velocities of flocculate particles.

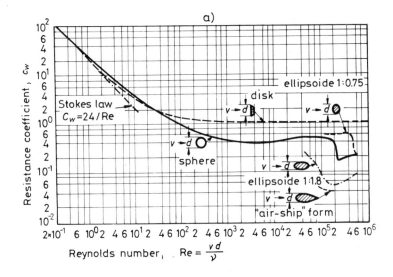

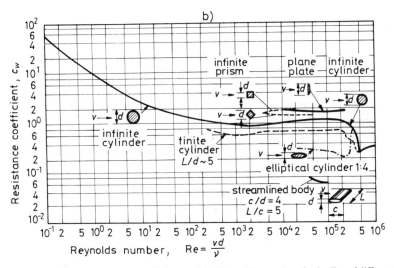

Fig. 2.24. Relationship between the drag coefficient and the Reynolds number, for bodies of different shapes

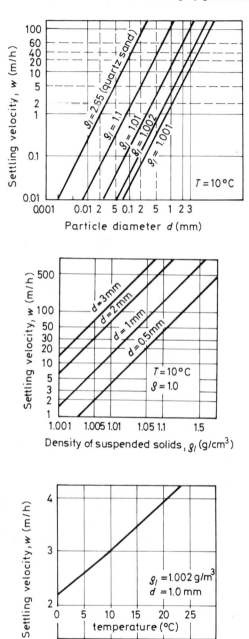

Fig. 2.25. Diagrams for calculating the settling velocity (after Oehler)

I. Mechanical treatment

Table 2.1/a. Settling velocity of sand and silt particles in stagnant water
(T = 10 °C) (after György 1974)

Description	Particle diameter (mm)	Settling velocity (mm/s)	Time required to cover a vertical distance of 1.0 m
Gravel	10	1000	1.0 s
Gravel from River Danube	2	200	5.0
Sand	1	100	10.0
	0.8	85	12.0
	0.6	63	16.0
	0.5	53	19.0
Sand	0.4	42	24.0
	0.3	32	31.0
	0.25	26.5	38.0
Fine sand	0.20	21.0	47.0
	0.15	15.0	1.1 min
	0.10	8.0	2.1
Fine sand	0.08	6.0	2.7
	0.06	3.88	4.3
	0.05	2.90	5.7
	0.05	2.90	5.7
	0.04	2.10	7.9
	0.03	1.30	12.9
Silt	0.02	0.62	26.9
Silt	0.015	0.35	47.6
	0.010	0.154	1.8 h
	0.008	0.098	2.8
	0.006	0.065	4.3
Fine silt	0.005	0.0385	5.7
	0.004	0.0247	11.2
Fine silt	0.003	0.0138	20.1
	0.002	0.0062	1.8 day
Clay	0.0015	0.0035	3.3
	0.0010	0.00154	7.5
	0.0006	0.00056	26.6
	0.00022	0.00006	192.7

Table 2.1/b. Settling velocities in stagnant waste water of 10 °C temperature
(after Fair et al. 1968)

Material	Specific weight (p/ml)	Settling velocity (mm/s)						
		Particle diameter (mm)						
		1.0	0.5	0.2	0.1	0.05	0.01	0.005
Quartz sand	2.65	140	72	23	6.7	1.7	0.083	0.017
Coal	1.5	42	21	7.2	2.1	0.42	0.022	0.004
Suspended solids of communal sewage	1.20	34	17	5	0.83	0.22	0.008	0.002

Finally, *Table 2.1/a* is presented (after the design standards of MÉLYÉPTERV, Hungary), showing the settling velocities of particles of different materials and sizes, indicating also the time required for covering settling distance of 1.0 m. The table is based on a literature review. Data in the table correspond to a 10 °C water temperature, from which the conversion to $T(°C)$ can be made by using eq (2.76).

Table 2.1/b presents the data of Fair for the settling velocities of various suspended particles in sewage water ($T = 10$ °C).

3 Sand traps

Sand traps belong to the category of mechanical treatment facilities of waste-water treatment plants. The purpose of building sand traps is to remove the particulate mineral content of raw waste waters entering the treatment plant with the highest possible efficiency, thus eliminating or decreasing problems that might occur during the technological process due to their presence (in order to decrease depositions in pipes, canals, structures, etc. — for example in digestors).

3.1 Settling velocity

In respect to the calculation of settling velocities of sand and other particulate matter, see Chapter 2. In the present chapter only some additional graphical guides will be presented.

Figure 3.1/a has been adopted from the classic work by Blunk (1933a). The figure indicates that, considering the settling velocities of sand and organic matter found in

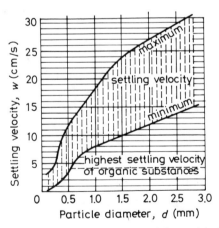

Fig. 3.1/a. Settling velocity of Rhine River sand as a function of the particle diameter (after Blunk 1933)

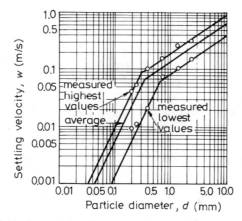

Fig. 3.1/b. Settling velocity of sand as a function of the particle diameter

Rhine river water, the efficient settling of mineral particles of diameters larger than 0.5 mm is expectable. Blunk proposes the curves of *Fig. 3.1/b* as an aid in design work. For the calculation of mean settling velocities (on the basis of the curve in the middle of the graph), the following empirical expressions are recommended:

$$w = 57\, d^2 \quad \text{(cm/sec)}; \quad d \leqq 0.33 \text{ mm} \tag{3.1a}$$

$$w = 13\, d^{0.66} \quad \text{(cm/s)}; \quad d \geqq 0.33 \text{ mm} . \tag{3.1b}$$

It is to be noted that eqs (2.36/a and b), proposed by Fair (in Fair et al. 1968), are of similar character. (*Table 2.1/b*, corresponding to sewage water of 10 °C temperature, was also suggested by Fair.) The data in *Table 3.1* (after Mostkow) are also of practical value.

Table 3.1. Settling velocities of sand particles
(after Mostkow, in Horváth 1984)

Particle size, d(mm)	0.01	0.02	0.05	0.1	0.2	0.5	1.0	1.5	2.0	5.0
Settling velocity w(cm/s)	0.005	0.02	0.13	0.55	1.8	5.0	10.7	16.3	19.0	30.0
w(m/h)	0.18	0.72	4.7	20	65	180	385	590	685	1080

Remark: Spherical particles; temperature range: $T = 10$—15 °C

3.2 Critical bottom-flow velocity

Various manuals and technical guides generally propose a throughflow velocity of
30 cm/s for the hydraulic design of sand traps. The smallest settleable particle
diameter may be taken as 0.1—0.2 mm, bearing in mind that particle size should be
characterized by a well-defined distribution function.

From the design point of view, the relationship between critical bottom velocity
and the particle size has a special importance. In *Fig. 3.2* the curves constructed by
Bogárdi (1972), on the basis of literature data, are presented, as modified by the data
of the author's own experiments. Curve No. 1 corresponds to the theoretically
obtainable suspension velocity. Curves 2—4 were adopted from Camp (1942b) and
refer to three different friction coefficients. Curve 5 was graphically determined by
Perry (in Bogárdi 1972) on the basis of data obtained from settling tanks in operation.
The relationship represented by curve 6 was derived by Bogárdi (1972) on the basis
of his laboratory experiments. These relationships refer mostly to settling facilities
and are applicable in the case of spherical particles. Finally, curves 7a and b were
defined by the author on the basis of prototype scale experiments with air-diffusion
sand traps in the treatment plant of Pécs, Hungary.

The equations of these curves are

$$v_k = 97.0 \, d^{0.77} \quad \text{for} \quad 0.01 < d \, (\text{mm}) < 0.23 \tag{3.2a}$$

and

$$v_k = 63.5 \, d^{0.5} \quad \text{for} \quad 0.23 < d \, (\text{mm}) < 2.5 \,, \tag{3.2b}$$

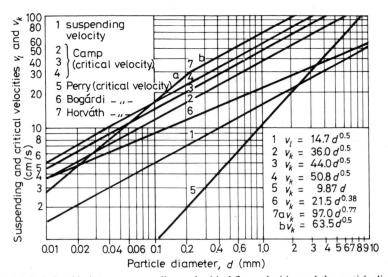

Fig. 3.2. Relationship between suspending and critical flow velocities and the particle diameter

respectively. It is worthwhile to note that the data published by Neighbor and Cooper (1965) well fit the relationships defined by the author. The figure also shows that a velocity of 16—28 cm/s corresponds to the particle diameter range 0.1—0.2 mm, matching well with the generally accepted 30 cm/s as the upper limit.

Scheme of the sand trap

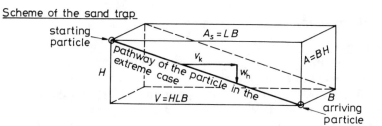

a) Laminar flow (Hazen's concept)

residence time: $t_c = \dfrac{V}{Q_v} = \dfrac{HLB}{HBv_k} = \dfrac{L}{v_k}$

settling time: $t_s = \dfrac{H}{w_h}$

in extreme case: $t_c = t_s;$ $\quad \dfrac{L}{v_k} = \dfrac{H}{w_h}$; $\dfrac{V}{Q_v} = \dfrac{H}{w_h}$

Surface load:
$$T_f = \frac{Q_v}{A_s} = w_h; \quad (k = \frac{w_h}{T_f} = 1)$$

Bottom area of the basin:
$$A_s = \frac{V}{H} = L B = \frac{Q_v}{w_h}$$

Length of the basin:
$$L = \frac{v_k}{w_h} H = \frac{Q_v}{B w_h}$$

b) Turbulent flow (concepts of Hazen and Dobbins-Camp)

Bottom area of the basin:
$$A_s = k \frac{Q_v}{w_h} = \frac{Q_v}{T_f}; \quad (k = \frac{wh}{T_f} > 1; \ T_f < w_h)$$

Length of the basin:
$$L = k \frac{Q_v}{B w_h} = \frac{Q_v}{B T_f}$$

Conditions:

steady flow, uniform velocity distribution, constant settling velocity (particulate matter)

Fig. 3.3. Basic expressions for designing sand traps with longitudinal throughflow

3.3 Sand traps with longitudinal throughflow

These sand traps belong to the category of classical structures. A characteristic example of this category is the Essen type sand trap.

Hydraulic principles

Sand traps of the longitudinal throughflow type can be designed on the basis of surface loading rates. The respective relationships are summarized in *Fig. 3.3* (theoretical principles will be further discussed in Chapter 4). Relationships under *a*), refer to laminar flow, according to the original approach by Hazen (in Ivicsics 1968). Under the effect of turbulence, settling velocities will change, as they also have vertical pulsation components. In order to take this effect into consideration, the Dobbins–Camp procedure will be presented here, that calculates bottom area A_s as follows (see Chapter 4 for more detail).

$$A_s = \frac{w_h}{T_f}\frac{Q_v}{w_h} = k\frac{Q_v}{w_h} = \frac{Q_v}{T_f}; \quad k>1, \qquad (3.3)$$

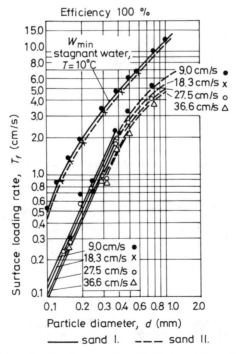

Fig. 3.4/a. Allowable surface loading rate of longitudinal sand traps (after Kalbskopf 1966)

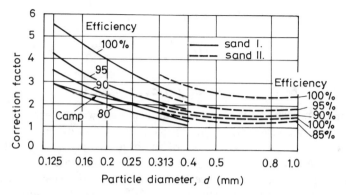

Fig. 3.4/b. Correction factors for taking the effects of turbulence into account in sand traps with longitudinal throughflow (after Kalbskopf 1966)

in which $k = w_h/T_f$ can be considered a correction factor, the value thereof can be obtained as a function of the intensity of turbulence. Kalbskopf (1966a) has further developed this procedure, extending the range of validity for sand traps of different types. His relationships are summarized in *Figs 3.4/a* and *3.4/b*. *Figure 3.4/a* shows the relationship between particle size and the allowable surface loading rate for sand traps of the longitudinal throughflow type. From *Fig. 3.4/b* the value of the correction coefficient $k = w_h/T_f$ can be obtained directly, for various particle sizes and at a given treatment efficiency. The effect of turbulence has to be taken into consideration above velocities of approximately $v_k > 5$ cm/s (a more accurate determination of this range can only be obtained on the basis of the Reynolds number).

Finally, the following two expressions can be used for designing sand traps in the case of laminar and turbulent flows, respectively

$$A_s = \frac{Q_v}{T_f} = LB; \qquad A = \frac{Q_v}{v_k} = HB. \tag{3.4a, b}$$

By fixing the value of one of the main dimensions, a certain degree of freedom in design is obtained; for example, by appropriately selecting the cross-sectional area of throughflow, a velocity $v_k = 30$ cm/s can be maintained at a nearly constant value, even in the case of changing Q_v values. Knowing Q_v and calculating T_f from the above expressions, the value of the two main dimensions can be obtained when a priori fixing the third one (e.g. the width of the basin).

For example: Considering eq (3.1a) the design relationship, the required length of the sand trap is obtained, for laminar flow, as:

$$L = \frac{Q_v}{Bw_h} = 1.75 \frac{Q_v}{Bd^2}; \qquad d \leq 0.33 \text{ mm} \tag{3.5}$$

and the necessary water depth is given as:

$$H = \frac{Q_v}{Bv_k}.$$

(3.6)

Determination of the mean throughflow velocity

In designing sand traps of the longitudinal throughflow type care should be taken to maintain the throughflow velocity at approx. 30 cm/s. One of the means of assuring this is to design the cross-sectional area of the basin in such a way as to maintain nearly constant flow velocities even at changing discharge rate Q_v. Thus, with the assumption of $v_k =$ const, the required cross-sectional area is obtained from the relationship $A = Q_v/v_k$. Several methods are available for the practical realization of this concept. For example: a trapezoidal shape of the cross-section geometry will assure nearly constant flow velocities. Another possibility is provided by designing parallel basin units — velocity-control channels — where the different units are put into operation at different flow rates, by adjusting the overflow weirs to the respective water depths.

Sand traps controlled by Parshall and Venturi flumes and by special weirs belong to a frequently used category of design. *Figure 3.5* shows the design of weir profiles for three different cross-sectional geometries; squared, triangular and parabolic (quasi-trapezoidal) shapes.

Parshall flumes or weirs controlling the water level should be installed at the outflow end of the sand trap. The condition of appropriate operation is to assure supercritical flow. In this case, the downstream water level will not affect the water level of the sand trap.

3.4 Sand traps with vertical throughflow

The best-known designs are the funnel-shaped and Blunk-type sand traps. In both cases, waste water first flows vertically downward in the basin, thus separating suspended mineral particles. In the case of the Blunk-type sand trap, nearly uniform mean flow velocities are assured by diversion cylinders placed along concentric circles. In this manner, with increasing Q_v more and more sectors will join the process, as is assured by the respectively set weir-edge heights. At low flows only the outer zone is in operation, then with the increase of Q_v more and more sectors will contribute to the discharge, while the water level will rise in the tank accordingly.

The relationship $A_s = Q_v/w_h$, defining the wetted bottom-surface area, remains valid also in the case of sand trape of vertical throughflow type. In this case, however $A_s = A$, the throughflow cross-sectional area, and refers to the area of the respective circular rings (i.e., those that contribute to discharge). Kalbskopf (1966a) demonstrated that the ratio w_h/T_f, correcting for the effect of turbulence, can be taken in this case as unity. Consequently, in sand traps of this type, those particles will settle

a) Rectangular cross-section

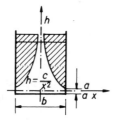

Weir profile: $h = \dfrac{v^2 b^2}{8g} \dfrac{1}{x^2}$; $\quad x = \dfrac{vb}{2\sqrt{2g}} \dfrac{1}{\sqrt{h}}$

$$a = \frac{v^2}{2g} \; ; \quad c = \frac{v^2 b^2}{8g}$$

Channel profile: $x = \dfrac{b}{2}$

b) Triangular cross-section

Weir profile: $x = \sqrt{2ph}$

Channel profile: $h = \dfrac{2v}{\sqrt{3pg}} x$

$$x = \frac{3pg}{2v} h \; ; \quad m = \frac{2v}{\sqrt{3pg}}$$

c) Parabolic cross-section

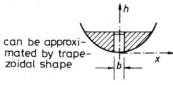

can be approxi-
mated by trape-
zoidal shape

Weir profile: $x = \dfrac{b}{2}$

Channel profile: $h = \dfrac{6v^2}{gb^2} x^2$

$$x = \frac{b\sqrt{gh}}{\sqrt{6}\,v}$$

v = flow velocity (approx. 0.3 m/s)
g = acceleration of gravity, 9.81 m/s^2
p = distance of focal point
    (equations do not include loss coefficient μ)

Fig. 3.5. Weir profiles for assuring constant mean flow velocity in sand traps with longitudinal throughflow (after Pallasch and Triebel 1967)

out the settling velocities of which are higher than the surface loading rate. The ratio w_h/T_f, representing the effect of turbulence, can be considered as unity, because turbulence will only displace the particles horizontally. In the vertical direction, however, pulsation might result, at the most, in a retarding effect. By selecting an appropriate basin depth, the range of settleable particles will not be altered by these effects. In practice, however, shock-like changes in flow velocities (in the case of the system not being able to fully eliminate changes of the loading rates), and the potentially present stagnant space might result in deviations from those predicted by the above theoretical formulas.

With respect to the more important basic design data, the following guide lines can be taken into consideration. The maximum vertical inflow velocity should not exceed

1.0 m/s. The length of submerged diversion cylinders must be at least 3.0 m, in order
to provide a sufficiently long pathway for the efficient settling of mineral particles.
Funnel-shaped settler bottoms should be designed with a slope exceeding 1.5:1.
This design will enable the pumping out of downward slipping deposited substances.
In order to prevent clogging, possibilities for air and water flushing must be en-
sured.

According to Blunk's original measurements, sand traps should be designed for a
surface load of 10 cm/s. The respective particle size, above which the sand becomes
settleable at this settling (or more accurately, suspending) velocity, is 0.4—0.6 mm.

According to more recent investigations, Blunk-type sand traps will assure good
operation at surface loading rates of 5—10 cm/s. It should be taken into consideration, however, that (according to Blunk's original concept) these settling tanks are
aimed mostly at the removal of coarser (d > 0.4 mm) sand and gravel particles and
especially at low discharge rates.

It is to be noted that several versions of the Blunk-type sand trap design exist. The
one proposed by Lohff (in Blunk 1933a), for example, uses a floating conical funnel,
which rises when the discharge increases, thus expanding the throughflow cross-
section and securing constant flow velocities.

3.5 Tangential sand traps

Geiger (1942) laid down the hydraulic principles of sand traps with tangential
inflow, and developed the design known as the Geiger sand trap. Although circular
and helical through-flow type designs were proposed by Hinderks, as early as in 1930,
the practical realization is tied up with Geiger's name.

Later, Kalbskopf (1966a) and Ostermann (1956) carried out significant research
work into the processes taking place in such sand traps, gaining more accurate
knowledge on the respective properties. This Chapter is based mostly on the works
of the above-mentioned authors.

Hydraulic principles

From the hydraulics of water courses, it is known that sediment deposition occurs
mostly on the inner side of river bends. This phenomenon is, however, not due to the
presence of lower flow velocities. Moreover, under certain circumstances, higher flow
velocities might occur at the inner side of such bends.

In the first approximation let us assume (along with Böss (1955) and Geiger (1942)
the vortex free flow of an ideal fluid in a curved channel section with rectangular
cross-section. According to the theory of potential flows

$$v = \frac{\text{const}_1}{r},$$
(3.7)

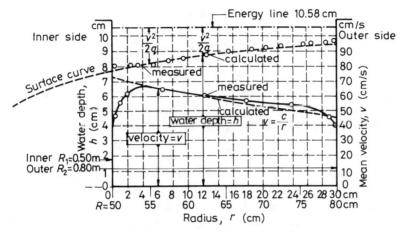

Fig. 3.6. Measured and calculated flow velocities and water depths in the bend of a rectangular channel
(after Geiger 1942)

where $\mathrm{const}_1 = C$ can be obtained from the basic expression for the definition of discharge Q, as

$$Q = \int_{R_1}^{R_2} v\, \mathrm{d}F = \int_{R_1}^{R_2} \frac{C}{r} h\, \mathrm{d}r \qquad (3.8)$$

taking the notations of *Fig. 3.6* also into account. Changing water depth h can be expressed as a function of bend radius r from Bernoulli's equation as:

$$\frac{v_0^2}{2g} + h_0 = \frac{v^2}{2g} + h = H; \quad h = H - \frac{v^2}{2g} = H - \frac{C^2}{r^2 2g}, \qquad (3.9)$$

and substituting into eq (3.8) one obtains:

$$Q = \int_{R_1}^{R_2} \frac{C}{r}\left(H - \frac{C^2}{r^2 2g}\right) \mathrm{d}r. \qquad (3.10)$$

Integrating eq (3.10),

$$Q = CH \ln \frac{R_2}{R_1} + \frac{C^3}{4g}\left(\frac{1}{R_2^2} - \frac{1}{R_1^2}\right), \qquad (3.11a)$$

whence the value of C can be obtained when H, R_1, R_2 and Q are known. Using the approximation $H = h_0$, one obtains that

$$C = \frac{Q}{h_0 \ln \dfrac{R_2}{R_1}}. \qquad (3.11b)$$

The distribution of water depths and flow velocities along the curved channel section is shown in *Fig. 3.6* as obtained from the above expressions and as measured in experiments. It is seen that velocity increases on approaching the inner side of the channel, the exception being the near-to-the-wall parts of the measured curve. The water surface is not characterized by the frequently assumed concave paraboloid, rotation-generated surface, but by a convex hyperbolic surface, resulting from Rankin's vortex motion. Moving away from the channel wall and considering a complete vortex, the following expression can be written for the inner core of this latter:

$$v = \mathrm{const}_2\, r .\qquad\qquad(3.12)$$

In this range the water surface corresponds to that of a rotating vessel, i.e., a paraboloid generated by rotation. Finally, the flow conditions of a structure with tangential water inflow will be such as is shown in *Fig. 3.7*. These conditions are characterized by: a) a water surface as discussed above, b) a velocity distribution as shown in the Figure, c) the development of transversal currents, and d) by the development of an interim separation zone according to the theorem of Helmholtz (in Bogárdi 1972).

From the operational viewpoint of sand traps, transversal currents have a specific significance, as they facilitate the transport of substances settled out in the middle

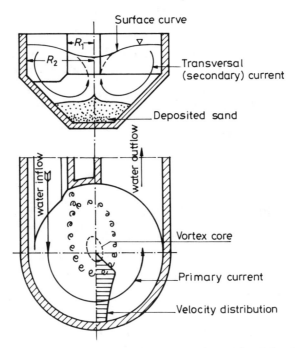

Fig. 3.7. Flow conditions in the Geiger type sand trap (after Geiger 1942)

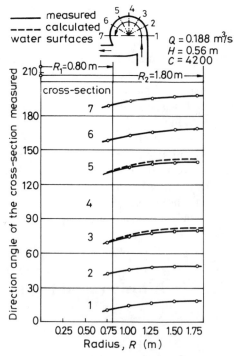

Fig. 3.8. Water surface curves in tangential sand traps (after Geiger 1942)

parts of the structure to the sand sump. Transversal currents superimposed on the circular flow result in a helical flow, similar to that of the currents in natural stream bends, mentioned above. Geiger (1942) explained the occurrence of transversal currents in sand traps as resulting from wall effects and from the transversal slope $I = v^2/gr = \text{Fr}$. It can be concluded that high v and small r values are needed for creating an appropriately high slope I.

Finally, *Fig. 3.8* is presented on the basis of Geiger's experiments, showing calculated and measured water surface curves corresponding to $Q = 0.188$ m³/s, $h_0 = = 0.546$ m, $R_1 = 0.8$, $R_2 = 1.8$ m and $H = 0.56$ m. The value of C was 4200 cm/s, or 4250 cm/s as calculated by using the approximate eq (3.11b).

Design data

The design inflow velocity, as proposed by Geiger, should be in the range of 0.75 m/s—1.0 m/s. The outflow velocity may not exceed 70—80 cm/s. The ratio of the sand trap volume to the inflowing sewage volume per minute should be between 25/l and 30/l. That is, for a maximum sewage discharge Q_v $t_c = 25$—30 s. The proposed average residence time should be 35—40 s. The volumetric design is then determined as $V = Q_v t_c$. Knowing the effective volume V, the main tank dimensions can be calculated by maintaining a diameter to volume ratio of 2 : 1. Considering a sand trap of length $L = 15$ m and of longitudinal throughflow type, with a mean flow velocity of 0.3 m/s, the required average residence time is obtained as $t_c = 15 : 0.3 = 50$ s.

On the basis of experiments carried out in circular-flow-type sand traps, Kalbskopf (1966a) found that with turbulent flow, pulsating velocities have a substantial effect on the settling velocity of suspended particles — similarly to the case of longitudinal throughflow type sand traps. It was also found that the hydraulic efficiency is about 50%, a rather unfavourable value.

In recent years, several new research results have been obtained in respect to the processes taking place in tangential sand traps, in the course of searches for designs with better removal efficiency (e.g., the research results of Ostermann 1967). Nevertheless, there are still many structures in operation that are inefficient and uncontrollable. Systematic studies could however, generally facilitate the improvement of the operational properties of inefficient circular sand traps (and of tangential grease and oil traps operating on the basis of the same principles).

3.6 Air-diffusion sand traps

Operation principles

Air diffusion sand traps operate on the principles of the mammoth pump effect. Air bubbles are introduced at a specified depth below the water surface, thus inducing currents in the plane of the basin cross-section, while the magnitude of flow velocities can be varied within wide ranges by adjusting the parameters of air injection (air discharge rate, injection depth, etc.). The flow induced in the plane of the basin cross-section, and especially in the critical lower bottom zone, plays an important role in the operation of the structure and in setting its removal efficiency.

This transversal flow will thus be referred to subsequently as the primary current. Due to the sewage discharges onto the structure, longitudinal currents will also develop, called the secondary currents. The latter are related to the residence time. The superimposing of transversal and longitudinal currents results in a helical flow that is regulated, to a greater or lesser extent, by the geometry of the basin. Transversal flow velocities are higher than the longitudinal ones; the latter thus play a secondary role only. An important conclusion to be drawn from this is that the treatment efficiency is practically independent, within certain limits that can be determined experimentally, of the discharge and/or the residence time.

A major advantage of air-diffusion sand traps is their flexible controllability. Namely, for given operational conditions, the flow velocities, and thus the removal efficiency of particulate mineral matter, can be varied within wide ranges. Moreover, the settling properties of organic matter can also be adjusted by varying the air-injection rate. Air-diffusion sand traps have accordingly been more widely employed in practice, during the past two decades.

Hydraulic conditions

In investigating the hydraulic properties of air-diffusion sand traps, the primary tasks are to define the flow pattern, velocity distribution and optimal basin geometry. The flow conditions of these sand traps can be analysed either on the basis of hydro-mechanical considerations or experimentally. By synthesizing the results obtained in these two ways, practically useful conclusions can be drawn and expressions can be derived.

Flow pattern and hydraulically favourable basin cross-section design

In the first approximation, the flow in an air-diffusion sand trap can be considered as a circular motion around a vortex with horizontal axis. In this vortex, the flowing medium makes, theoretically, a potential circular motion. It is known from hydromechanics that the velocity distribution in a vortex line can be approximated with the linear relationship $v = k_1 r$ (similarly to eq 3.12), on analogy with the rotating motion of solid bodies. At the same time, for the close vicinity of the vortex line, the hyperbolic relationship $v = k_2(r)$ similarly to eq (3.7) becomes valid. The obvious idea then arises, that the velocity ranges described by the above two expressions be approximated by a single relationship.

It is easily conceivable that the two expressions concerned are special cases of the expression

$$v = \frac{ar}{b + r^2},$$
(3.13)

since for $b \gg r^2$, $k_1 = a/b$ and for $b \ll r^2$, $k_2 = a$. A further task is to interpret parameters a and b hydraulically. Two conditional equations are needed for this. One of the conditions is created by defining the intersection point, the common point, of the two curves, to which maximum velocity v_{max}, and radius r_m correspond. Thus

$$v_{max} = k_1 r_m = \frac{k_2}{r_m},$$

that yields the hydraulic interpretation of parameter b as:

$$\frac{k_2}{k_1} = b = r_m^2.$$
(3.14a)

The hydraulic interpretation of parameter a can be derived from eq (3.13), as the second condition: when velocity v_{max} corresponds to radius r_m, then a can be expressed from eq (3.13) as

$$a = 2r_m v_{max}.$$
(3.14b)

Knowing a and b the hydraulically interpreted form of eq (3.13) can be written in a dimensionless form as

$$\frac{v}{v_{max}} = \frac{2}{\dfrac{r}{r_m} + \dfrac{r_m}{r}}.$$
(3.15)

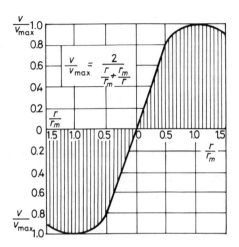

Fig. 3.9/a. Dimensionless theoretical transversal velocity distribution in air-diffusion sand traps

It can be seen that the dimensionless velocity ratio v/v_{max} depends solely on the dimensionless variable r/r_m. This relationship is illustrated by *Fig. 3.9/a. Figure 3.9/b* presents an example of a practical application for a sand trap of diameter $D_H = 4$ m, with $v_{max} = 30$ cm/s and $r_m = 1.9$ m. It must be emphasized, however, that due to the approximate character of the above considerations, the thus derived velocity distributions are also approximate. Nevertheless, as was revealed by experimental investigations, the theoretical velocity distribution approximates the actual conditions well, and allows the drawing of qualitative conclusions as well as the making of approximate design calculations.

In deriving eq (3.15) it was implicitly assumed that no diversion plates or other obstacles disturb the development of circular motion in the basin. Thus, the velocity distribution obtainable in the above manner will only be valid in those sections of the sand trap where this condition of the absence of disturbance prevails. It follows

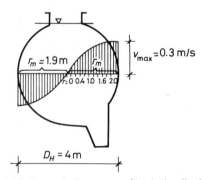

Fig. 3.9/b. Example for transversal velocity distribution

directly from the above hydromechanical considerations that the most favourable cross-sectional geometry of air-diffusion sand traps is the circular design. This conclusion has also been supported experimentally. The ideal cross-sectional geometry can also be approximated by an appropriate polygon.

There are some practical cases where the depth of the structure is limited due to one reason or another. Deviations from the circular geometry can be allowed in such cases with the condition that the depth to width ratio of the basin remains smaller than unity. In practice, it is expedient that this ratio should be between 3/4 and 4/4.

Relationships between variables defining the hydraulic conditions and the bottom flow velocity

In this section, the role of some other important variables characterizing the hydraulic conditions of air-diffusion sand traps will be investigated and empirical relationships between them presented on the basis of experimental results. The experiments were carried out in a circular structure of diameter D_H, as shown in *Fig. 3.10*.

Bottom-flow velocities, which play a decisive role in the operation of air-diffusion sand traps, are affected by the following variables: the structure's geometric design, the depth of air injection, the discharge rate of air injection, the buoyant velocities of bubbles, the physical properties of the fluid, and the acceleration of gravity. On the basis of experimental results the following empirical expressions, which permit actual design calculations, have been derived:

$$v = 236\, h_r^{0.78}\, Q_{air}^{0.7}.\tag{3.16}$$

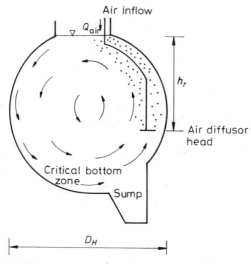

Fig. 3.10. Schematic cross-section design of air-diffusion sand traps

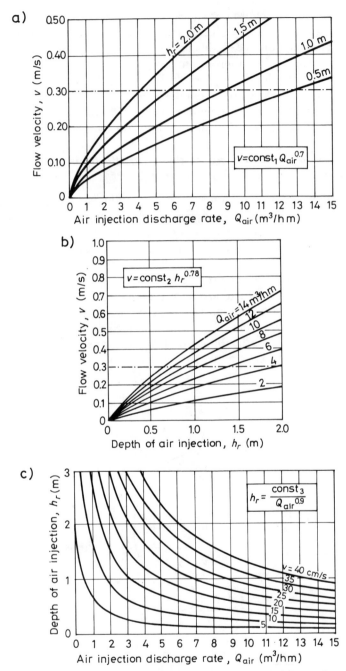

Fig. 3.11. Design guide for relating bottom zone flow velocity to air discharge and to the depth of air injection

Table 3.2/a. Air discharges suggested for air injection sand traps (after Hartmann 1966)

Air injection depth (m)	Minimum air flow rate needed for trapping sand $(m^3/h \cdot m)$	Maximum preliminary aeration rate (without obstructing sand trapping operation) $(m^3/h \cdot m)$
1.5	12.5—15.0	30
2.0	11.0—14.5	29
2.5	10.5—14.0	28
3.0	10.5—14.0	28
4.0	10.0—13.5	25

In order to facilitate practical application, a diagram (for use as a design aid) has been constructed on the basis of the above expression. *Figure 3.11a* presents the v versus Q_{air} relationships for air-injection depths of 0.5, 1.0, 1.5 and 2.0 m. *Figure 3.11b* shows the v versus h_r relationship for the 2 $m^3/h \cdot m < Q_{air} < 14$ $m^3/h \cdot m$ parameter interval. Finally, in *Fig. 3.11c* relationship h_r versus Q_{air} is shown. In these figures a dashed line marks the locations of the critical velocity, 30 cm/s. The curves of the figures were derived on the basis of eq (3.16) and correspond to a basin diameter $D_H = 2$ m.

According to the investigations by Hartmann (1966) the required air-discharge rates can be estimated on the basis of *Table 3.2/a* (the data correspond to a Heilbronn-type sand trap). For air-diffusion sand traps with a cross-sectional area of 1.5 m^2, Kalbskopf (1966a) suggests the use of the data in *Table 3.2/b*.

The relationship between air injection, electric power requirement and the effective cross-sectional area is shown in *Fig. 3.12*. The curves were defined by Kalbskopf (1966a) with the condition that the bottom-flow velocity should be at least 30 cm/s (even in the presence of surface-active substances). In the case of deep air injection, the guideline value of air discharge is 1.5 $Nm^3/h \cdot$ basin m^3.

Table 3.2/b. Experimental results with an air injection sand trap of 1.5 m^2 cross-sectional area (after Kalbskopf 1966a)

Sand particle diameter (mm)	Surface loading rate (mm/s) settling (removal) rate			Settling velocity of sand in stagnant water (mm/s)
	100%	90%	85%	
0.125	1.0	1.7	2.2	8.6
0.16	1.6	2.6	3.3	13.5
0.20	3.0	4.8	5.9	19.0
0.25	5.5	8.2	10.0	15.0
0.315	8.8	12.3	14.7	35.0

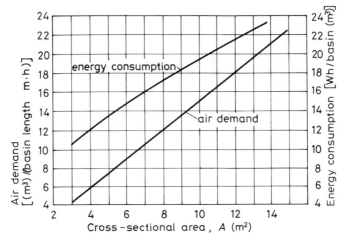

Fig. 3.12. Air injection discharge and energy requirement of air diffusion sand traps

Conversion to systems of differing dimensions

In conjunction with investigations carried out with air-diffusion sand traps, we developed a method for the determination of the conversion coefficient of the air-injection discharge. Considering the special operational conditions of air-diffusion sand traps, the following simplified approximate relationship can be applied (Horváth 1966b):

$$\lambda Q_{air} = \frac{2\lambda^2}{1+\lambda^{-1/2}} = \frac{2\lambda^{5/2}}{\lambda^{1/2}+1}, \tag{3.17a}$$

where $\lambda = 1'/1''$ is the geometrical conversion factor, that is the inverted scale, and

$$\lambda Q_{air} = Q'_{air}/Q''_{air}$$

is the conversion factor of air discharges.

Relating the air discharge to the unit basin length in both systems,

$$Q_{air} = \frac{2\lambda}{1+\lambda^{-1/2}} = \frac{2\lambda^{3/2}}{\lambda^{1/2}+1}. \tag{3.17b}$$

Figure 3.13 shows the curves described by eqs (3.17a and b), together with the curve definiable on the basis of Froude's law (curve No. 1). The latter obviously deviates from curves 2 and 3, since the changes of λQ_{air}, obtainable on the basis of Froude's law, do not appropriately reflect the effects of bubble motion.

The above relationships allow the conversion of the design data, obtained from *Figs 3.11a—c* for a structure of basin diameter $D_H = 2.0$ m, to those corresponding to a structure with an arbitrarily selected basin diameter.

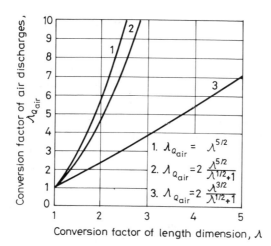

Fig. 3.13. Conversion of air discharges for sand traps of different dimensions

Design principles

According to our experience, air-diffusion sand traps can be applied economically for treatment plants of capacities larger than 15,000—20,000 population equivalent. In selecting the type of the structures local conditions should naturally also be taken into account. The main dimensions of the sand traps should be designed on the basis of the maximum values of hourly averages. In determining the effective cross-sectional area, the basic condition should be that the axial mean flow velocity must not exceed 15—20 cm/s. Thus bottom flow velocities are essentially determined by the transversal flow component induced by air injection.

The calculation of the volume of the structure can be made on the basis of the average residence time; the value thereof — when related to the maximum of hourly averages — is 2.5—3.0 min (in a rainy period), or in the case of preliminary aeration, 10—12 min (in a dry period). It is generally advisable to take the hydraulic efficiency (the ratio of stagnancy free volume to the total volume) also into account. This is done in such a way as to divide the volume, obtained by the above calculation, by the hydraulic efficiency.

The value of the latter should be determined on the basis of laboratory experiments. In the absence of such experiments, 80—90% efficiency can be considered. Knowing the geometric dimensions of the structure, the required air-injection rate can be calculated. In determining the total area of the holes in the air diffusor heads, an air outlet velocity of approx. 1.0 m/s can be considered.

3.7 Rotor-aerated sand traps

As is known, air-diffusion sand traps have been developed on analogy with air-injection aerator basins in the early fifties. Later the obvious idea arose that rotor-aerated sand traps can be developed following the example of horizontal axis type rotor-aerating devices (such as the Kessener system). On this conceptual basis, scale-model studies were started in 1965 to determine the operational parameters with which horizontal axis type rotors could be used for aerating sand traps. The results of these experiments can be summarized as follows (Horváth 1966b):

a) It was found that in sand traps equipped with rotor aerators, favourable flow conditions can be achieved. By adjusting the submersion depth of the rotors, the flow velocities in the critical near-bottom zone can be well controlled.

b) The main operational parameters suggested were the following: speed of the rotor 60—80 r.p.m.; diameter of the rotor 60—70 cm; the angle between the horizontal plane and the diversion plates above the rotor 30—40°; and the vertical distance between the lower edge of the diversion plate and the water level 5—10 cm.

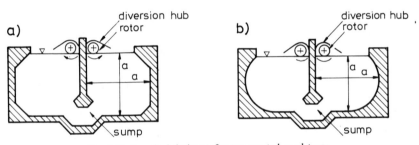

Fig. 3.14. Practical designs of rotor-aerated sand traps

c) On the basis of hydraulic investigations, proposals were made for the design of the cross-sectional geometry of the structure. Two alternative solutions are shown in *Fig. 3.14a* and *b*, respectively.

3.8 The stratification of sand

The investigations by Ostermann (1967) revealed that mineral particles exhibit a marked stratification in flowing sewage water. The type of stratification developing in rectangular and trapezoidal channels leading to tangential sand traps is illustrated by *Fig. 3.15/a* and *3.15/b*. The vertical distribution of the sand concentration is very characteristic. Most of the sand will be concentrated in the lower 10 cm layer. It is worthwhile to note that the highest concentrations do not occur at the channel walls,

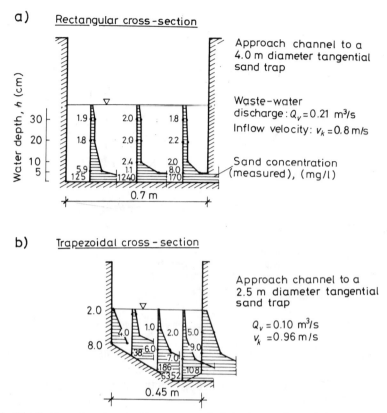

a)   Rectangular cross-section

Water depth, h (cm)

Approach channel to a
4.0 m diameter tangential
sand trap

Waste-water
discharge: $Q_v = 0.21$ m³/s
Inflow velocity: $v_k = 0.8$ m/s

Sand concentration
(measured), (mg/l)

0.7 m

b)   Trapezoidal cross-section

Approach channel to a
2.5 m diameter tangential
sand trap

$Q_v = 0.10$ m³/s
$v_k = 0.96$ m/s

0.45 m

Fig. 3.15/a. Distribution of sand content in an approach channel to sand trap (after Ostermann 1967)

but along the axis. This phenomenon might mostly be due to the transversal currents induced by wall friction (similar effects are encountered in the tangential sand traps, as discussed previously). The wedge shaped depositions in the longitudinal directions of channels formed with flat bottoms are due to the same cause.

The vertical distribution of sand concentration in open channels was described mathematically by Einstein (in Bogárdi 1972), and this can also be applied to sand traps. Assuming a steady, uniform and turbulent flow the relationship is written in the following form:

$$\frac{C_y}{C_a} = \left[\frac{h-y}{y} \frac{a}{h-a} \right]^z , \qquad (3.18)$$

where h is the water depth; $a = 0.1\,h$; y is the distance above the channel bottom; $z = \dfrac{w}{0.4u}$ (w is the settling velocity, u is the shear velocity; for steady uniform flow, $u = \sqrt{gRI}$, in which R is the hydraulic radius and I is the slope); C_a and C_y are

I. Mechanical treatment

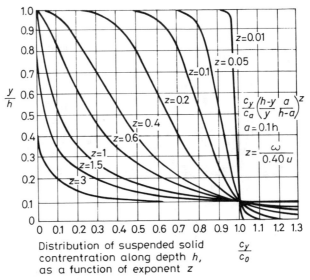

$$\frac{c_y}{c_a}\left(\frac{h-y}{y}\cdot\frac{a}{h-a}\right)^z$$

$a=0.1h$

$$z=\frac{\omega}{0.40\,u}$$

Distribution of suspended solid contrentration along depth h, as a function of exponent z

$\dfrac{c_y}{c_0}$

Fig. 3.15/b. In-depth distribution of suspended solid concentration (after Einstein, in Bogárdi 1972)

the suspended solid concentrations at a characteristic point a above the bottom, and at a point of variable distance y above the bottom, respectively.

Equation (3.18) is presented in graphical form in *Fig. 3.15*. In a given case, knowing w, R and I, the exponent

$$z = \frac{w}{0.4u}$$

can be calculated (where 0.4 is the Prandtl constant). The concentration ratio C_y/C_a can be obtained from the respective curves as a function of the y/h value. When a certain characteristic concentration C_a is known, the value of any selected C_y can be expressed. Efficient settling can only be expected with smaller turbulence, that is with higher z values. This is well represented by the pattern of the curves. With the increasing of turbulence — i.e. with the decreasing of z — the curves converge to a vertical line $C_y/C_a=1$, indicating a uniform concentration distribution.

3.9 Longitudinal particle distribution

In sand traps of the longitudinal throughflow type, the longitudinal distribution of particle size is a characteristic feature of the hydraulic conditions. In this context, *Fig. 3.16a* and *b* are presented after the measurement results by Hartmann (1966). *Figure 3.16a* shows the particle-size distribution along the longitudinal dimension of the sand

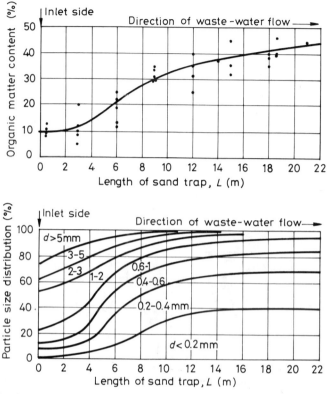

Fig. 3.16. Longitudinal distribution of settled sand particles in a sand trap with longitudinal throughflow

trap, for different particle size fractions. The *Figure* indicates that sand particles of diameter $d>1$—2 mm can be removed with 90% efficiency in a structure of 12 m length. In contrast, the fraction of $d<0.2$ mm can not be efficiently settled out, not even in a basin of 22 m length. According to *Fig. 3.16b* the quantity of settled organic substances also increases with the length of the structure.

Figure 3.17 illustrates the sand-particle distribution along the bottom of a tangential sand trap of 4.0 m diameter. The bulk of the large quantity of sand will — in this given case — settle out until flow reaches the section corresponding to an angle of 140°.

3.10 Efficiency of sand traps

The suspended material content of sewage water entering a sand trap becomes separated and split up due to the mechanical effects of the flow and participates in a settling process as a function of the flow velocities and of the specific weight of the

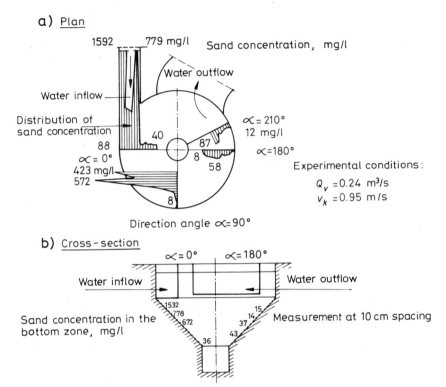

Fig. 3.17. Distribution of sand content in the bottom zone of a tangential sand trap (after Ostermann 1967)

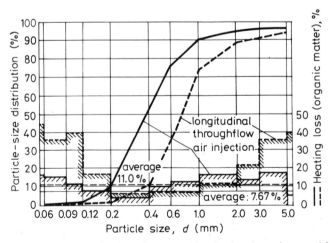

Fig. 3.18/a. Particle-size distribution of deposited sand and heating loss (organic matter) in sand traps of longitudinal throughflow and air injection sand traps, respectively (after Pöpel and Hartmann 1958)

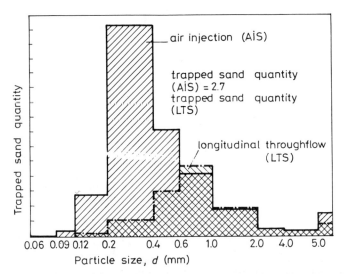

Fig. 3.18/b. Comparison of deposited sand fractions in sand traps of longitudinal throughflow and with air injection

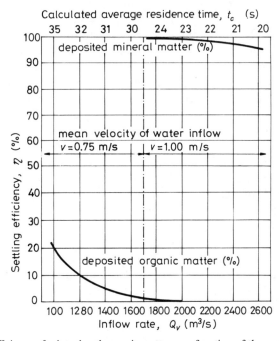

Fig. 3.19. Settling efficiency of mineral and organic matter as a function of the average residence time in Geiger-type sand trap (after Geiger 1942)

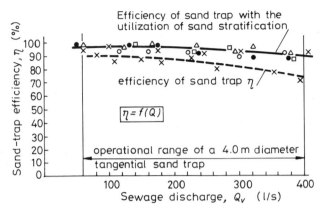

Fig. 3.20. Efficiency of tangential sand trap as a function of the sewage discharge (after Ostermann 1967)

particulate matter. The objective of operating sand traps is, in all cases, to remove the mineral fraction of the suspended solid content with the highest possible efficiency, while assuring the further transport of the organic fraction with the effluent of the sand trap. It should be emphasized already in advance that this goal is never completely achieved, since the suspended solids in waste waters cover a very wide range of specific weight and contain organic and inorganic substances of widely varying origin. Thus, in respect of the settleability, a gradual transition represents the various fractions. Obviously, in some special cases, there might be such fractions present (e.g., rough sand and gravel from stormwater washoff loads) that will settle out much faster than the organic substances, thus allowing an efficient separation. In a general practical case, however, the trapped particles will always contain a certain percentage of organic matter, even in the best sandtraps, while the effluent of the sand trap always contains some particulate matter of mineral origin.

Figures 3.18/a and *3.18/b* allow a comparison of the removal efficiencies of sand traps with longitudinal throughflow and of those with diffuse air injection, for organic and inorganic matter. It can be seen that air-diffusion sand traps are more efficient in removing the finer fractions of sand.

Relationships characterizing the efficiency of Geiger sand traps are shown in *Fig. 3.19*, as a function of residence time t_c and of the inflowing discharge. The variation of the efficiency of tangential sand traps is demonstrated by *Fig. 3.20*, as a function of the waste-water discharge Q. The difference between the two curves shown in the Figure is due to the elimination of sand stratification (by mixing) in one of the approach channels. This latter technique resulted in a smaller removal efficiency. Thus it can be concluded that the stratification of sand has a favourable effect on the removal efficiency of sand traps.

4. Settling facilities

4.1 Kinematic characterization of settling particles

From hydraulical and technological points of view, settling facilities consist of the following main parts: a) inlet space (primary chamber); b) settling space; c) outlet space (secondary chamber); and d) sludge space. Kinematic investigations of the settling process are generally focussed on the actual settling space, while also taking the inlet and outlet structural parts into account, because of their important role in forming the flow pattern.

One of the possible ways of characterizing settling facilities kinematically is to describe the pathway of motion of settling particles. The pathway of a settling particle is defined by the sum of the vectors of throughflow velocity v and settling velocity w. Taking the greatly varying designs of settling facilities into consideration, there are very many possible theoretical solutions to this problem. Some of the technologically most important of these will be reviewed in this Chapter.

$A(a)$ Settling basin with longitudinal throughflow. $Q_v =$ const; $v =$ const. The flow is of steady state (local and convective accelerations are zero, hydrostatic head distribution). Settling velocity $w =$ const. Particulate, non flocculate matter is considered. The pathway of the settling particle is a straight line as shown by *Fig. 4.1/a*. With laminar flow and uniform velocity distribution, one may assume that the settling process takes place as if it were occurring in settling cylinders (under hydrostatic conditions) that travel with the flow in the structure.

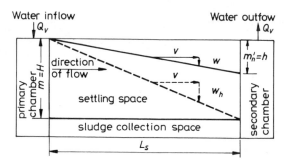

Fig. 4.1/a. Settling of particulate matter in a settling basin with longitudinal throughflow

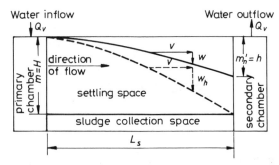

Fig. 4.1/b. Settling of flocculate matter of low concentration in a settling basin with longitudinal through-flow

A(b) If $w \neq$ const (but the rest of the above conditions remains unchanged) then the pathway of particle motion will such as shown in *Fig. 4.1/b*. This case refers to the settling of flocculate particles. The increasing steepness of the curve indicates that w increases with time (due, for example, to the increasing size of flocs).

B(a) Settling basins with radial throughflow, $Q_v =$ const, $v \neq$ const. Steady, abruptly changing flow (local acceleration is zero but convective acceleration cannot be disregarded; pressure distribution is other than hydrostatic). For $w =$ const the equation describing the pathway of settling particles can be written, with the symbols used in *Fig. 4.1/c*, as

$$ y = \frac{2\pi w}{Q_v}\left[R_0 T_0 x + (T_0 - R_0 \tan \gamma)\frac{x^2}{2} - \tan \gamma \frac{x^3}{3}\right]. \tag{4.1} $$

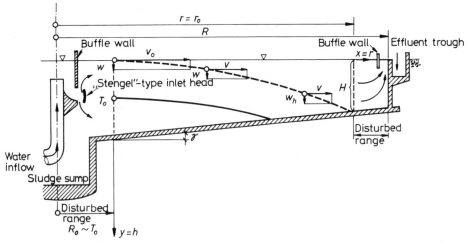

Fig. 4.1/c. Settling of particulate matter in a settling basin with radial throughflow

$B(b)$ If $w \neq$ const, then the mathematical formulation of kinematic properties becomes an even more difficult task.

$B(c)$ In settling basins of the vertical throughflow type, both the streamlines and the particle pathways will — in an ideal case — be vertical.

In practice, however, most situations will differ, to a smaller or larger extent, from the above-described ideal cases; e.g., the flow is turbulent and non-steady ($Q_v \neq$ const), velocity distributions are other than uniform, and stagnant spaces occur.

4.2 Hazen's concept

The original concept of Hazen defines the basic design equation for settling basins with longitudinal throughflow, subjected to the following initial conditions: a) the settling velocity of particles in flowing media is the same as in stagnant fluid; b) the concentration of suspended matter is constant at any point of any cross-sections that are perpendicular to the direction of flow that is the concentration distribution is uniform even in the inflow section; c) flow is steady, the velocity distribution is uniform and the direction of flow is horizontal.

Considering the scheme of a longitudinal settling basin as shown in *Fig. 4.1/a*, together with the above assumptions, the following expressions characterize the movement of a particle that starts from a surface point of the inflow cross-section and arrives at the bottom point of the outflow cross-section:

Time of settling, $t_s = H/w$;
Residence (throughflow) time:

$$t_c = L/v = LBH/vBH = V/Q_v;$$

Volume of the settling basin: $V = LBH$;
Rate of flow $Q_v = BHv$.

If the settling time equals the time needed for covering the useful settling distance (i.e., when $t_s = t_c$), then $H/w = L/v = V/Q_v$, whence the effective settling length is

$$L = H \frac{v}{w} \tag{4.2}$$

and the settling velocity, corresponding to the extreme case, is

$$w = \frac{H}{L} v = \frac{Q_v H}{V} = \frac{Q_v}{BL} = \frac{Q_v}{A_s} = T_f \quad \left(\frac{m^3/h}{m^2}\right) = \left(\frac{m}{h}\right), \tag{4.3}$$

where $w = w_h$ the settling (extreme) velocity of the smallest particle to be settled out; $v = v_k$ mean flow velocity in the structure (see *Fig. 3.3*).

Relationship $T_f = Q_v/A_s$ defines the surface loading rate; T_f means the height of a water column that is loaded onto the water surface, $A_s = BL$ due to discharge Q_v, during a unit period of time.

The basic conclusion that can be drawn from Hazen's concept is that $w_h = T_f$, on the basis of which the required basin area A_s and settling depth L are obtained as

$$A_s = \frac{Q_v}{w_h} = \frac{Q_v}{T_f}; \quad L = \frac{Q_v}{w_h B} = \frac{Q_v}{T_f B}. \tag{4.4}$$

Let us consider a case, as an example, when $w = w_h$ can be calculated on the basis of the Stokes law:

$$w = \frac{gd^2}{18v} \frac{\gamma_1 - \gamma}{\gamma} = 0.545 \frac{d^2}{v} (\gamma_1 - 1),$$

and by substituting into eq (4.4):

$$A_s = \frac{Q_v}{w_h} = 1.834 \frac{v}{\gamma_1 - 1} \frac{Q_v}{d^2}. \tag{4.5}$$

The effective length $L = A_s/B$ or effective radius $R \approx \sqrt{A_s/\pi}$ of longitudinal and radial settling basins can thus be respectively obtained as:

$$L = 1.834 \frac{v}{\gamma_1 - 1} \frac{Q_v}{B d^2},$$

and

$$R = 0.765 \sqrt{\frac{v}{\gamma_1 - 1} \frac{Q_v}{d^2}}. \tag{4.6a—b}$$

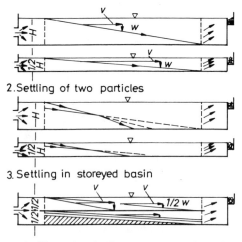

1. Settling of a particle in an ideal basin with longitudinal throughflow

2. Settling of two particles

3. Settling in storeyed basin

Fig. 4.1/d. Examples illustrating the Hazen's concept (after Pallasch and Triebel 1967)

It should, however, be emphasized also here that the design settling velocity w_h should be determined by actual measurements.

It is seen from the above relationships that the values of A_s, L and T are not affected by water depth (settling depth) H. Similarly, neither residence time t_c play a role in the design. Consequently, the design parameters are settling velocity w_h and surface loading rate T_f. The condition that water depth H and residence time t_c do not play a role in the design can be easily understood on the basis of the following relationships:

$$T_f = \frac{Q_v}{A_s} = \frac{V}{A_s t_c} = \frac{H}{t_c}. \tag{4.7}$$

Namely: at a constant discharge rate $Q_v = \text{const}$, residence time t_c increases proportionally with the increase of water depth H, while the throughflow velocity decreases. This means that a longer residence time will be available for covering a greater settling depth (examples illustrating this relationship are shown in *Fig. 4.1/d*).

Hazen's original concept, first published in 1904, can be used — in accordance with initial conditions listed under item *a*) — for designing settling facilities with longitudinal or radial throughflow. The possibilities of expanding the concept of designing settling facilities on the basis of the surface-loading rate to facilities with vertical throughflow as well as to particulate and flocculate substances, will be discussed later.

4.3 The method of Dobbins and Camp

With respect to the design of settling facilities, several experiments were made to investigate the turbulent character of flow. The role of turbulence can be examined by applying a correction factor to the settling velocity, or by considering turbulence-induced dispersion (mixing) effects explicitly. In this latter context, the method of Dobbins and Camp (1944) will be presented below.

Dobbins described the changes in suspended solids concentrations over time, by the following differential equation, relying on the transport of suspended solids by turbulent dispersion:

$$\frac{\partial C}{\partial t} = \varepsilon \frac{\partial^2 C}{\partial y^2} + w \frac{\partial C}{\partial y}, \tag{4.8}$$

where y is the vertical distance, i.e., the vertical coordinate of a point in the water body, and ε is a coefficient of mixing (turbulent dispersion).

Assuming a parabolic velocity distribution in the settling basin, Camp (1942b) derived the following relationships to express the dispersion coefficient as a function of shear stress τ, density ϱ and water depth H:

$$\varepsilon = 0.075 H \sqrt{\frac{\tau}{\varrho}} \tag{4.9}$$

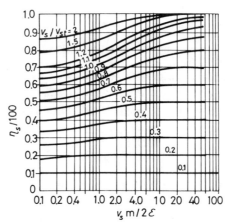

Fig. 4.2. Consideration of the effect of turbulence in a settling basin with horizontal throughflow (after Camp) $(v_s/v_{st} = w/w_t; \ v_s m/2\Sigma = wH/2\Sigma)$

$$\frac{wH}{2\varepsilon} \approx 122\,\frac{w}{v} = 122\,\frac{w_f}{v}\,\frac{w}{w_t} = 122\,\frac{H}{L}\,\frac{w}{w_t}, \qquad (4.10)$$

where $wH/2\varepsilon$ is a dimensionless number expressing the extent of turbulence. On the basis of these relationships, Camp constructed the series of curves shown in *Fig. 4.2*, where parameter w/w_t is the ratio of settling velocity in stagnant water to that in turbulent flow. The efficiency of settling is shown on the vertical axis. The curves represent solutions of eq (4.8), in a dimensionless approach. According to Camp, the ratio L/H ranges, in practice, from 2/3—200, and w/w_t from 0.1—2.0. Accordingly, the value of $wH/2\varepsilon$ — as determined by eq (4.10) — will fall in the interval 0.06—81.

In practice, the method of Dobbins and Camp can be applied for designing settling facilities as follows. Assuming that the values of variables w and v are known, $wH/2\varepsilon$ (the value of the variable on the horizontal axis) can be calculated by eq (4.10). Then, for a desired removal efficiency $(\eta_s/100)$ the series of curves yields the ratio w/w_t as required. Knowing w, the value of w_t can be obtained directly. The required length of the settling basin is then calculated as $L = Hv/w_t$. Thus, knowing the value of w_t, the effect of turbulence can be determined.

Hazen's concept, based on the notion of surface loading rate, is obviously in accordance with the design method of Dobbins and Camp. Namely, in this latter case, the settling velocity w of a particle in laminar flow should be substituted by the turbulent settling velocity w_t and thus:

$$A_s = \frac{Q}{w_t} = \frac{w}{w_t}\,\frac{Q}{w} = \frac{Q}{T_f}. \qquad (4.11)$$

Actually, this means that Hazen's original $T_f = Q/w$ relationship is modified by multiplying with a correction factor w/w_t, a parameter to be obtained from the curves of *Fig. 4.2.* The value of w_t can thus be considered as the surface loading rate allowable in the case of turbulent flow ($w_t = T_f$).

The conditions for applying the method of Dobbins and Camp are that the concentration of suspended solids be uniform in the full inflow cross-section and that mixing coefficient ε should be constant over the entire settling space. It is also assumed that the above-mentioned parabolic velocity distribution prevails, a condition better representing practical cases than the assumption of a linear distribution.

This method is primarily applicable for particulate (non flocculate) substances (e.g., sand) and for settling structures of the horizontal throughflow type. Design parameters for structures of a different type (e.g., sand traps) can be obtained from the works of Kalbskopf (1966a) (see Chapter 3). It is to be noted that according to the results of experiments by Wiegmann and Müller-Neuhaus (1952/53), the mean velocity of throughflow in circular settling basins should be a maximum of 20 times higher than the design settling velocity in order to assure that the settling efficiency must not be substantially effected by turbulence.

4.4 Surface loading rate and residence time

The design of settling structures is based principally on two basic parameters: the surface loading rate and the residence time. If the process of thickening is also to be considered, then a third parameter, the specific surface area, should be determined as well.

There is a strong relationship (as defined by eq 4.7) between surface loading rate T_f and the calculated residence time t_c. For a given basin depth, the surface loading rate and the residence time are, theoretically, two equally important design parameters, which are independent of the properties of the substance to be settled. In practice, however, the properties of the substance to be settled should also be taken into consideration. This is the reason why the surface loading rate is considered the main design parameter in the case of particulate matter, while the residence time becomes the basic design parameter in the case of flocculate substances of low concentration. Particulate matter settles out with a constant velocity in contrast to flocculate particles, silt flocs, the settling velocity of which changes substantially with the process of flocculation, i.e., with changing of the volume of flocs. This, however, does not mean that surface loading rates can be disregarded in the case of flocculate substances. In the special case of waste waters containing both particulate and flocculate suspended solids (e.g., in communal sewage waters), both parameters should be taken into account equally.

In settling facilities of the vertical throughflow type (for example in Dortmund-type basins), the surface loading rate becomes the design parameter, regardless of the character of the suspended solids. Namely, in this case the loading rate T_f also

Table 4.1/a. Proposed throughflow times (after Pallasch and Triebel 1967)

	Calculated throughflow time (h)			
	primary settling		final settling	
	longitudinal flow	radial flow	longitudinal flow	radial flow
Mechanical treatment only	1.7	1.7—2.5	—	—
Flocculation with chemical dosage	0.5	0.5—0.8	1.5	1.5—2.0
For trickling filters	1.5	1.5—2.3	1.5	1.5—2.0
For treatment of activated sludge	0.5	0.5—0.8	1.7	1.7—2.7

means the mean velocity of upward flow, as contrasted to settling basins of the horizontal throughflow type, in which only the units of T_f will be of a velocity character. In Dortmund basins, residence time t_c will be higher than 2.0 h, when one assumes a surface loading rate $T_f = 1.5$ m/h and a bottom slope of 60°. Consequently, designs on the basis of T_f and t_c are in conformity with the requirements of practical cases.

The proposed average residence times and allowable surface loading rates are summarized in *Tables 4.1/a* and *4.1/b*, for the most widely used settling structures, and for communal sewage water. In approximate calculations, the settling velocity of a particle of 0.05 mm diameter can be considered, according to Fair (in Fair et al. 1968), as 0.76 m/h (i.e., $T_f \leq 0.76$ m/h). With respect to the role of residence time, the already classical figures (*Figs 4.3/a, 4.3/b* and *4.3/c*) can be referred to. The joint effect of residence time and surface loading rate is demonstrated by *Fig. 4.4*, on the basis of the laboratory investigations by Pflanz (1966). The nearly horizontal pattern of the efficiency curves indicates that the effect of T_f is smaller than that of t_c. (If T_f has no effect on the efficiency at all, then horizontal lines would characterize the relationship.)

There are some further considerations concerning the distribution of residence times. Methods known as throughflow waves or throughflow curves are the most

Table 4.1/b. Proposed maximum surface loading rates

Treatment system	Surface loading (m/h)				
	primary settling basin		final settling basin		
	longitudinal flow	radial flow	longitudinal flow	radial flow	Dortmund basin
For mechanical treatment only	1.3	1.3—0.8	—	—	—
For flocculation with chemicals	4.0	4.0—2.5	1.5	1.5—1.0	1.5
For trickling filters	1.3	1.3—0.8	1.5	1.5—1.0	1.5
For treatment of activated sludge	4.0	4.0—2.5	1.2	1.2—0.7	1.2

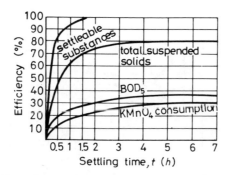

Fig. 4.3/a. Effect of residence time on the settling efficiency of municipal sewage (after Sierp 1953)

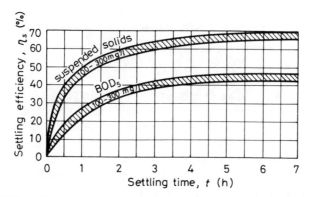

Fig. 4.3/b. Effect of residence time on the primary settling efficiency of municipal sewage

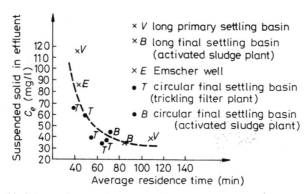

Fig. 4.3/c. Relationship between the actual average residence time and the suspended solid content of the effluent (after Schmidt–Bregas 1958)

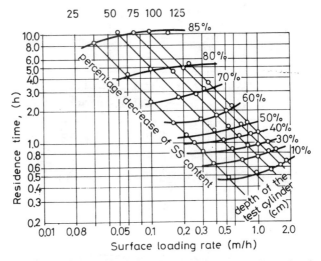

Fig. 4.4. Settling efficiency as a function of the residence time and the surface loading rate

widely applied. We would like to mention only that the above-mentioned methods allow the determination of the size of dead or sluggishly flowing tank-volume percentages (i.e., the extent of hydraulic short circuiting), and thus the hydraulic efficiency of a structure.

The changes of hydraulic efficiency as a function of the basin dimension are illustrated by *Figs 4.5a* and *b*, on the basis of the experimental results of Schmidt–Bregas (1958), for Dorr-type settling basins with longitudinal and radial throughflow, respectively. On the basis of these figures the following conclusions can be drawn.

a) In settling structures of the longitudinal throughflow type, the hydraulic efficiency increases with decrease of the ratio basin depth/basin length. This finding is in

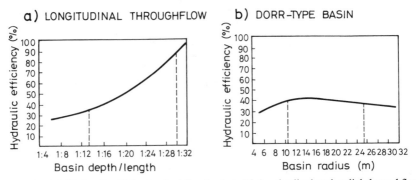

Fig. 4.5. Hydraulic efficiency of settling basins with longitudinal and radial throughflow

accordance with the principle of the so-called "pipe reactors", characterized by plug flow or piston flow.

b) In circular Dorr-type settling structures, the hydraulic efficiency is the highest, in the given case, at basin diameters of 12—16 m, but decreases with further increase of the diameter. The absolute value of hydraulic efficiency is 30—50 per cent. However, this does not mean that Dorr-type settling structures must be designed with diameters of 12—16 m. The above-mentioned hydraulic efficiency means a purely hydraulic qualification, whereas in practice the entire technology should be taken into account.

c) Both the longitudinal and the radial basins have horizontal.throughflow, but their hydraulic properties are different. The difference is due to the fact that in Dorr-type basins the velocity decreases with increasing cross-sectional area, thus leading to a higher potential for the development of dead-flow zones. Consequently, ideal flow conditions, characteristic of pipe reactors, can only be maintained with difficulty.

d) It follows from the above considerations that settling structures of longitudinal throughflow type are hydraulically more favourable than those of the Dorr type, since the optimal conditions of settling would occur in ideal pipe reactors.

4.5 Design aids for dimensioning primary and secondary settling structures

The relationship between the suspended solid loads of secondary settling basins and the suspended solid content of their effluent is illustrated (after Pflanz 1969), in *Figs 4.6/a* and *b*, for two characteristic temperature ranges. Surface loading rate T_f is obtained by dividing the suspended solid load onto the settling basin by the sludge concentration of the aerated mixed liqueur discharged into the basin. *Figures 4.6/c* and *d* show the relationship between the sludge concentration of the aerating basin

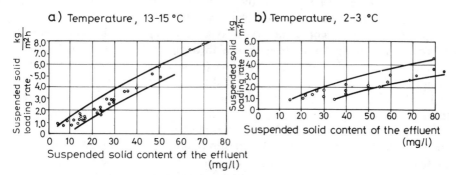

Fig. 4.6/a–b. Relationship between the suspended solid loading rate of longitudinal settling basins and the suspended solid content of the effluent

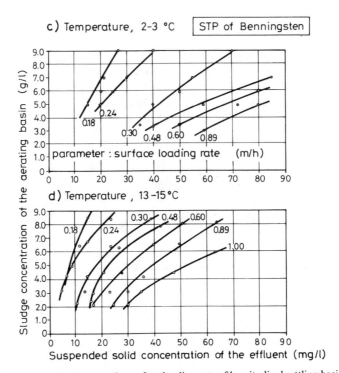

Fig. 4.6/c–d. Relationship between the surface loading rate of longitudinal settling basins and the suspended-solid content of the effluent

and the surface loading rate T_f, or the suspended solid concentration of the treated effluent, for the same experiments. These experiments were carried out in the final settling basin of the sewage-treatment plant at Benningsen (Germany). The main parameters of this structure are: $V=100$ m³; $A_s=84$ m²; $H=1.2$ m; $H/L=1:17.5$; $t_c=3$ h; the specific discharge over the weirs $=8.33$ m³/h·m; the Froude number $Fr=5.1\times10^{-7}$; and the hydraulic efficiency $\eta=100\ t_m/t_c=80\%$.

Figure 4.7(a—c) presents nomograms to serve as guides to the design of settling basins of the longitudinal, radial and vertical throughflow types. The users should first calculate the required basin volume $(V=t_c Q)$ as a function of which the main dimensions of the structures can be obtained from the respective graphs. For basins with longitudinal throughflow, it is desirable to calculate the maximum length of the settling structure $(L_{max}=t_c v_{max};\ v_{max}\approx1.0$ cm/s$=36$ m/h; $L_{max}=36\ t_c$, m). Nevertheless, the length of the actual structure should be less than L_{max}.

In *Fig. 4.8* a design guide for settling basins of horizontal and vertical throughflow is presented which takes the activated sludge concentration of the aerating basin also into consideration. A more recent method for designing the final settling basins of activated sludge-type treatment plants was elaborated by Merkel (1974); this also

a) BASIN WITH LONGITUDINAL THROUGHFLOW   b) BASIN WITH RADIAL THROUGHFLOW

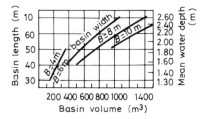 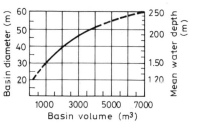

c) DORTMUND-TYPE BASIN

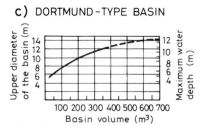

Fig. 4.7. Design guides for settling basins (after Pallasch and Triebel 1967)

takes the relationship between the aerating basin and the settling tank, as well as the treble role of the latter, into consideration. This treble role consists of settling (phase separation), thickening and sludge storage. Merkel's design-guide nomograms are shown in *Figs 4.9/a* and *4.9/b* for settling basins with horizontal and vertical through-flow. On the vertical axis of these graphs, the sludge concentration of the aerating basin is shown, as determined by 30 min settling experiments or by the Mohlmann

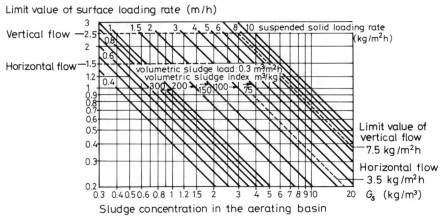

Fig. 4.8. Design guide for the settling basins of activated sludge plants

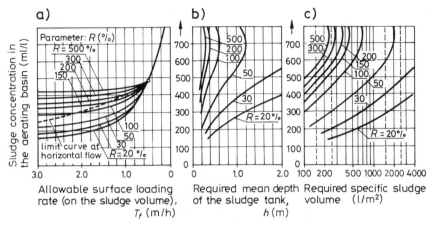

Fig. 4.9/a. Design guide for the final settling basins of activated-sludge plants with horizontal throughflow

index (ml/l) [sludge volume index sludge concentration (ml/g) · (g/l) = (ml/l)]. On the horizontal axis of the graphs, the following variables are plotted: a) allowable surface loading rate as calculated for the sludge storage space below the settling space (m); b) height of the sludge level (m); c) quantity of sludge (l/m²) [dry material content of the final settling tank (kg) times the sludge volume index (ml/g) divided by the area of the settling tank (m²)]. The parameter indicated by the curves is the recirculation rate in %. To calculate the total effective depth of the settling basin, an additional water depth of approx. 0.5 m should be added to the sludge level height. Dashed lines with arrows in *Fig. 4.9/a* correspond to the solutions for actual examples.

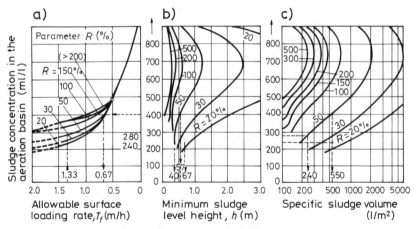

Fig. 4.9/b. Design guide for the final settling basin and thickener of activated-sludge plants with vertical throughflow (after Merkel 1974)

4.6 Design of settling basins of the "Uniflow" type

There exist settling facilities of special design, among which the "Uniflow"-type systems (also used in Hungary) will be presented below. The theoretical and experimental basis of these structures was elaborated by Dallas (1958). A characteristic feature of Uniflow systems is the higher bottom slope, as compared to that of basins with longitudinal throughflow, and the application of more transversal diversion troughs. The special design offers favourable hydraulic properties such as: a) effective energy dissipation at the inlet in order to reduce turbulence; b) more uniform velocity distribution; c) hydraulic stability due to converging streamlines; d) decrease of secondary currents (e.g., density flows and wind-induced currents); e) faster equalization of temperature changes; f) favourable conditions for sludge collection.

The nomograms for aiding design are presented in *Figs 4.10/a, 4.10/b* and *4.10/c*. (It is to be noted that the original American units are shown on the graphs, but conversion factors to metric units are also included.) These diagrams facilitate design in the 0.1—10.0 MGD range (1.0 MGD, i.e., one million gallons per day = 3785 m^3/day). To make use of this design aid, the following parameters should be selected experimentally: surface loading rate; width of basin; the length to width ratio of the basin; the ratio of throughflow to settling velocities and the specific discharge over the weirs.

4.7 Water inflow and outflow

It is known that the operation and efficiency of settling basins depends greatly on the design of the inlet and outlet facilities. Many alternative solutions are available for designers, from which the one best suited to the given conditions should be selected.

The correct design of the inlet structure is especially important. The kinetic energy of inflowing water, through the approach channel, should be abruptly decreased, via local impact losses and by creating turbulence, in order to eliminate the propagation of turbulent motion into the settling space. Some of the more widely used designs are the following: a) inflow over weirs; b) inflow through screens or perforated plates; c) the Geiger design; and e) the Stuttgart design. Inlet structure designs of the Geiger, Stengel and Stuttgart types are shown in *Fig. 4.11*. According to Pöpel and Weidner (1963), it is desirable to facilitate the development of a multiple "water cushion" at the obstructing element, in order better to dissipate the kinetic energy of flow. This was the concept behind the Stuttgart design, where a dual change of flow direction is enforced by a diversion plate and the wall of the basin itself.

Numerous observations related to the design of inlet structures have been made during trial operations of different sewage-treatment plants. Several constructional faults were identified; the inaccurate installation of sewage-water inlet elements, such as disks, could be blamed in many instances. Inaccurate construction may cause disturbances to the flow.

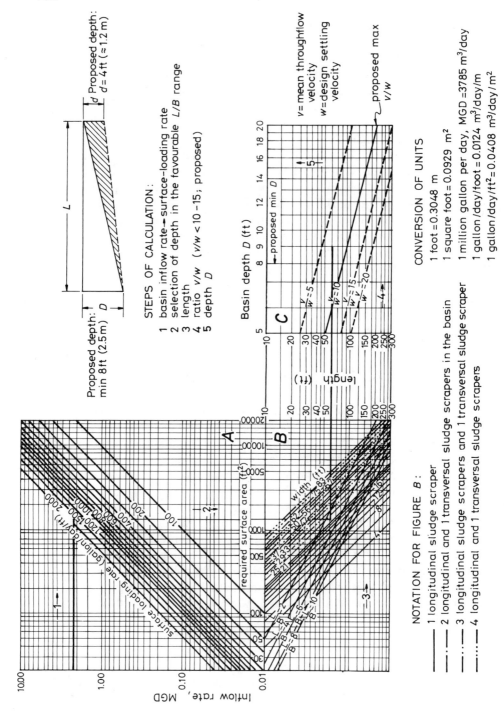

STEPS OF CALCULATION:

1 basin inflow rate → surface-loading rate
2 selection of depth in the favourable L/B range
3 length
4 ratio v/w (v/w < 10 –15; proposed)
5 depth D

Proposed depth:
d = 4 ft (≈1.2 m)

Proposed depth:
min 8 ft (2.5 m) D

v = mean throughflow velocity
w = design settling velocity

proposed max
v/w

CONVERSION OF UNITS

1 foot = 0.3048 m
1 square foot = 0.0929 m²
1 million gallon per day, MGD = 3785 m³/day
1 gallon/day/foot = 0.0124 m³/day/m
1 gallon/day/ft² = 0.0408 m³/day/m²

NOTATION FOR FIGURE B:

——— 1 longitudinal sludge scraper
—·— 2 longitudinal and 1 transversal sludge scrapers in the basin
—··— 3 longitudinal sludge scrapers and 1 transversal sludge scraper
····· 4 longitudinal and 1 transversal sludge scrapers

Fig. 4.10/a. Design guide for settling basins of "Uniflow" type (after Dallas 1958)

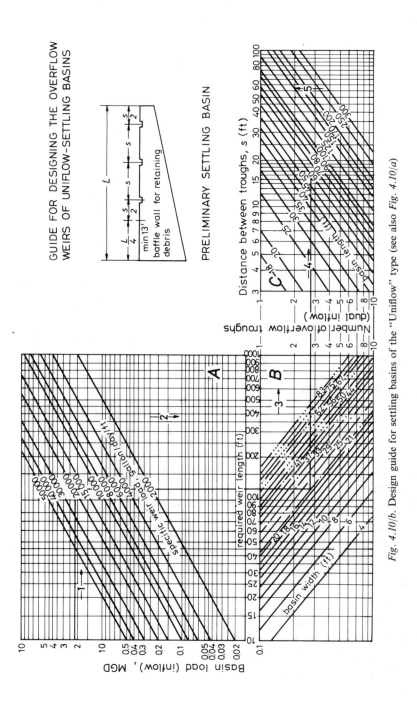

GUIDE FOR DESIGNING THE OVERFLOW
WEIRS OF UNIFLOW-SETTLING BASINS

PRELIMINARY SETTLING BASIN

Fig. 4.10/b. Design guide for settling basins of the "Uniflow" type (see also *Fig. 4.10/(a)*)

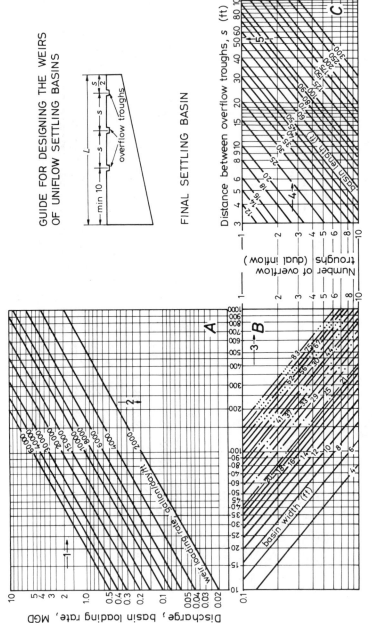

Fig. 4.10/c. Design guide for settling basins of "Uniflow" type (see also *Fig. 4.10/a*)

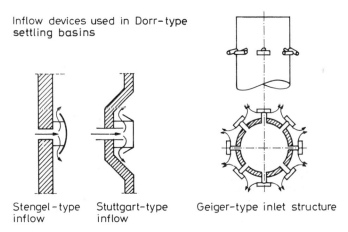

Fig. 4.11. Inflow arrangements used in settling basins

For example: the inaccurate — asymmetrical — installation of Geiger tubes or Stengel disks may cause one-sided loading of settling basins. In Dorr-type basins, deviation of the axis of the concrete cylinder that supports the Stengel disks from the vertical position can be the cause of one-sided loading. The same problem might also occur in Dortmund-type structures.

Inaccuracies of the above type might cause the development of considerable dead spaces. Generalizing, it may be stated that any deviation from symmetry (structural; water inlet or outlet; loading, water quality; physical or chemical) might cause one-sided loading, development of dead spaces and hydraulic jumps. On the one hand, this might result in a decrease of the effective volume and, on the other, the development of higher than desirable flow velocities. *Figure 4.12* shows some faulty structural solutions, with their effects on the flow conditions. The distorted scale of this figure serves to illustrate the phenomena better.

Some remarks on the role of outflow conditions should also be made. Generally speaking, this problem can easily be solved by installing appropriate weirs at the structures' upper edge, together with diversion troughs. More detailed analysis of the settling process and the flow conditions will, however, reveal the importance of selecting an optimal solution — a none too simple task in some cases. In order to assure the optimum hydraulic and settling efficiencies, water outflow might be facilitated by overflow troughs placed at different locations at the water surface (for example: in the case of circular structures, these troughs can be placed radially, concentrically or both). Such considerations resulted in the development of special structures. One of these special solutions is the Uniflow system, mentioned above, that can equally well be applied to rectangular or circular designs.

Studying the relevant literature, one might arrive at the conclusion that opinions on the effect of the specific loading rates of overflow structures are rather contradic-

OBLIQUELY INSTALLED "STENGEL" DISKS (DORR SYSTEM)

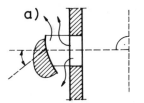

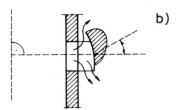

Most of the inflow is diverted
towards the water surface, while
dead spaces might occur in the
lower part of the basin

Most of the flow is diverted
towards the bottom of the basin,
dead volumes might occur in the
upper part, sludge might be
scoured out of the slurry sump

OBLIQUELY INSTALLED CONCRETE CYLINDER AND DIVERSION PLATE
(DORTMUND SYSTEM)

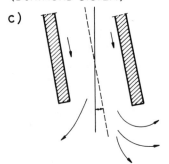

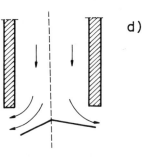

One-sided loading is caused by the asymmetric inflow

Fig. 4.12. Some commonly occurring faults of the inlet structures of settling basins

tory. Burdich (1964), in analysing the literature on Dortmund-type structures, found
that the effect of the edge-loading rate is usually over-estimated. This author considered the factor of correctly installing the overflow trough as being of greater
importance. Nevertheless, from the hydraulic point of view, the arrangement of
troughs and the selection of edge-loading rates, should be harmonized. In a narrower
range of load variations, however, the arrangement of the troughs will be more
important. The selection of the optimum rate of edge loading must, however, be based
on the due and simultaneous consideration of both the hydraulic and the settling
properties. Thus, for example, different optimum edge-loading rates will be obtained
for the different combinations of settling facilities (of longitudinal throughflow, Dorr,
or Dortmund type, etc.) and their associated primary and final settling basins.
Literature data suggest that edge-loading rates generally vary in the range of

2—40 m^3/m · h. Obviously, much higher rates are allowable in primary settling tanks than in the final settling facilities.

Some of the more important basic design data for water inflow and outflow are as follows. The flow velocity of sewage water in the approach channel should be in the range of 30—100 cm/s. If the velocity of throughflow across the settling structure is around 1.0 cm/s, then the respective energy dissipation rate should be one 30th or one 100th. At the inflow into the settling structure, the desirable velocity range would be 30—50 cm/s; this will allow the elimination of harmful depositions in the channel, while at the same time necessitating a smaller energy dissipation. In the case of flocculate substances, the edge-loading rate of water diversion troughs should — according to Imhoff (1966) — be in the range of 3.0—5.0 m^3/h · m, though other authors advocate that rates of 20—30 m^3/h · m will also allow of reliable operation. Schmidt–Bregas (1958) proposed 35 m^3/h · m as the edge-loading rate for the overflow trough installed at the end of a settling basin of longitudinal throughflow type. In designing the weir edges of structures serving for the settling of activated sludge, Anderson (1945) also considered the site for installing these weirs. He proposed 25 m^3/h · m as the edge-loading rate for troughs placed in the upward flowing zone, while 35 m^3/h · m was recommended for those in the horizontal flow zone. Maximum rates of 36.0 m^3/h · m and 108.0 m^3/h · m for primary settling basins and 18.0 m^3/h · m and 36.0 m^3/h · m for final settling tanks were suggested by von der Emde (1964) for dry and rainy season sewage flows, respectively.

4.8 Hydraulic design of the structural parts of settling facilities

The structural parts, such as water inflow and outlet structures, of settling basins need specific hydraulic design considerations.

Figure 4.13 shows, as a first example, the theoretical scheme for calculating the head loss caused by the submerged-cylinder regulated water-inlet structures, in cir-

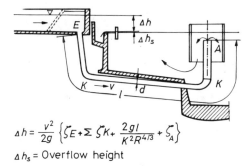

$$\Delta h = \frac{v^2}{2g} \left\{ \zeta_E + \Sigma \zeta_K + \frac{2gl}{K^2 R^{4/3}} + \zeta_A \right\}$$

Δh_s = Overflow height

Fig. 4.13. Hydraulic design of the submerged cylinder-type inflow structure of radial settling basins

cular settling basins, after the work of Groche (1964). *Figure 4.14* shows, as the second example, the structural design and the respective discharge calculation formulas of some inlet structures. In *Fig. 4.15*, a method for calculating the characteristic points of the surface curve in circular overflow troughs is presented, as the third example, for some possible hydraulic settings, together with the respective expressions. The design in this case follows the principles of open-channel hydraulics.

Finally, in *Fig. 4.16* a nomogram serving for the calculation of discharge over V-cog-shaped weir edges is presented. Further examples can be found in the well-known book by Fair, Geyer and Okun (1968).

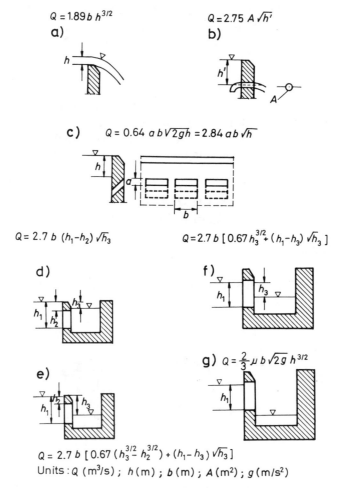

Fig. 4.14. Arrangements of outlet structures and the relevant discharge formulas
(after Müller-Neuhaus 1953)

a) $h_o = h_N \left\{ \sqrt{2 \left(\dfrac{h_k}{h_N}\right)^3 + \left(1 - \dfrac{I\,1}{3h_N}\right)^2} - \dfrac{2I}{3} \dfrac{1}{h_N} \right\}$

b) $h_N = h_k$ (water surface marked with dashed line):

$h_o = h_k \left\{ \sqrt{2 + \left(1 - \dfrac{I}{3} \dfrac{1}{h_k}\right)^2} - \dfrac{2I}{3} \dfrac{1}{h_k} \right\}$

c) For collection trough with horizontal bottom ($I=0$):

$h_o = h_N \sqrt{2 \left(\dfrac{h_k}{h_N}\right)^3 + 1}$

d) $I = 0$; $h_N = h_k$

$h_o = h_k \sqrt{3}$

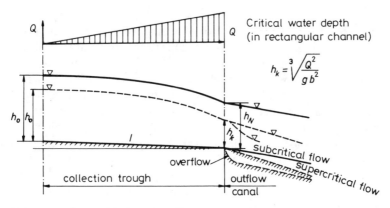

Critical water depth
(in rectangular channel)

$h_k = \sqrt[3]{\dfrac{Q^2}{g\,b^2}}$

Units: $Q\ (\mathrm{m^3/s})$; $b\ (\mathrm{m})$; $g\ (\mathrm{m/s^2})$

Fig. 4.15. Calculation of water surface curves occurring in circular overflow troughs

4.9 Flow velocities, flow pattern and stability

In applying settling structures, the flow velocity and thus the sediment-load carrying capacity of sewage water is decreased in order to facilitate the gravitational separation of suspended contaminating substances. It follows from this statement that both the absolute magnitude of flow velocities and their distribution are very important variables when discussing the operational properties of settling structures. There

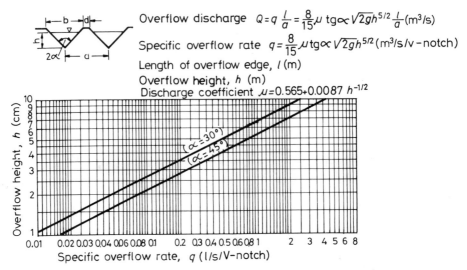

Overflow discharge $Q = q \, \dfrac{l}{a} = \dfrac{8}{15} \mu \, tg\alpha \sqrt{2g} h^{5/2} \dfrac{l}{a} \, (m^3/s)$

Specific overflow rate $q = \dfrac{8}{15} \mu \, tg\alpha \sqrt{2g} h^{5/2} \, (m^3/s/v\text{-notch})$

Length of overflow edge, l (m)

Overflow height, h (m)

Discharge coefficient $\mu = 0.565 + 0.0087 \, h^{-1/2}$

Fig. 4.16. Guide for the hydraulic design of V-notch overflow weirs (after Pöpel and Weidner 1963)

are, however, severe problems related to measuring techniques, as the absolute values of flow velocities are generally rather small. Due to these difficulties of direct measurement, some other methods, such as the use of various dies and tracers, have been developed for determining the throughflow properties of settling structures.

Standards prescribe the allowable flow velocities for different settling structures. The Hungarian standard defines a 1—2 cm/s limit velocity, as that at which no resuspension of already deposited substances occurs. The allowable maximum flow velocities in primary and final settling tanks are 4—5 cm/s and 2—3 cm/s, respectively, regarding the flow velocity in the vicinity of inflow, and not in the bottom zone. In the vicinity of a water inlet, the kinetic energy of water is decreased, thus offering the possibility for the development of turbulent motion, a rather unfavourable phenomenon in settling structures.

Due to the occurrence of stagnant, dead volumes and hydraulic short-circuiting, the flow velocities determined on the basis of geometric dimensions (i.e., cross-sectional area) can be considered only as guide values. The flow properties of settling facilities are rather complex and cannot be fully taken into consideration with idealized assumptions.

Purely hydraulic considerations were mentioned above. Nevertheless, it is unconditionally required that the relationship between hydraulic and settling properties be simultaneously taken into account in judging the efficiency of settling facilities. It is known that settling, flocculation and coagulation processes depend on the flow conditions. Flocs of different size coagulate to a certain extent also in standing waters, but reaction rates increase with the increase of turbulence to a certain limit value. Above this limit value, a further increase of turbulence will rather separate the flocs, i.e., it will hinder the flocculation process. Eventually, there might be a state of equilibrium, when the splitting up and generation of flocs becomes balanced.

The design principles of settling basins usually require the elimination of turbulence and/or that the calculations include the effect of turbulence. Solely from the viewpoint of settling, laminar flow is considered favourable, facilitating the unobstructed settling of particles; at the same time, however, turbulence might also be favourable as it increases the rate of flocculation.

As is known, the Reynolds number provides a measure for characterizing laminar and turbulent flows. The critical Reynolds number is that which marks the limits between laminar and turbulent flow. For settling structures, Re_{crit} is taken as the value corresponding to pipe flow:

$$Re_{crit} = \frac{vd}{v} = 2320 .$$

Introducing the hydraulic radius $R = d/4$ one obtains that

$$Re_{crit} = \frac{vR}{v} = 580 .$$

In prototype settling structures, the Re number drops, almost without exception, above the critical value, indicating turbulent-flow conditions.

There are, however, some critical remarks to be made regarding the above, generally accepted, statements.

a) The use of the critical Reynolds number of pipelines for designing settling structures can be misleading as the two systems are both geometrically and hydraulically different from each other.

b) The mean flow velocity, a parameter needed in calculating the Re number, can only be less exactly defined for settling structures due to the greatly varying pattern of flow in these structures.

c) According to Groche (1964), the hydraulic radius, replacing the diameter in the Reynolds formula, cannot be considered a physically based parameter, since it corresponds to a combined shape and not to a distance.

It follows from the above considerations that the Re_{crit} value as used in the design of settling facilities needs to be reviewed. There are such settling facilities (for example, long settling basins with small cross-sections of the longitudinal throughflow type) where the disputed critical Re value might be correct, whereas in most practical cases this value will be misleading. The practical justification of this statement is that in most settling basins there will be both laminar and turbulent flows, in a proportion that depends on the type of the structure.

It can thus be readily concluded that it would be desirable to separate different flow domains on the basis of throughflow experiments. In this manner, the conditions corresponding to ideal pipe reactors or to tank reactors can be defined and, moreover, the extent of hydraulic short-circuiting determined.

8*

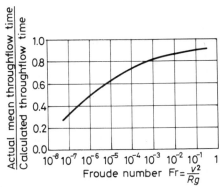

Fig. 4.17. Hydraulic efficiency of settling basin as a function of the Froude number (after Camp 1953)

It should not be said, however, that the determination of the laminar and turbulent flow domains is superfluous. But from the viewpoint of measuring techniques, it is more favourable to determine the distribution of residence times than to measure flow velocities. It could be a subject of further research to determine the relationship between flow conditions evaluated in these two ways.

In addition to determining the character of flow, the stability of flow is another parameter of main concern. The stability of flow means, generally, the capability of the flow to regain its original pattern following some disturbance, or after cessation thereof. The measure of stability is usually the Froude number

$$Fr = \frac{v^2}{gR},$$

mostly in situations dominated by gravitational forces. In the case of settling structures, however, the effect of the Reynolds number can also be significant. It follows that the use of the Froude number alone cannot provide a measure of the flow stability.

Camp (1953) investigated the effect of stability conditions on the hydraulic efficiency, as defined by the "throughflow wave" method. He found that hydraulic efficiency increases with increasing Froude numbers (stability increases; *Fig. 4.17*). This author did not consider the effect of the Re number. Schmidt–Bregas (1958) obtained similar results, with some exceptions. Groche (1964) criticized the findings of Camp and Schmidt and Bregas, emphasizing that the Re number also changes with the change of the Froude number, a condition that must not be neglected. It can thus be readily concluded that hydraulic efficiency is not so closely related to the Froude number as the relevant literature claims. According to our own opinion, the actual conditions could be better described by relating hydraulic efficiency to both Re and Fr in a dimensionless relationship, thereby implicitly considering the forces of inertia, gravity and friction equally well. This dimensionless relationship could also facilitate the modelling of the respective processes.

An empirical possibility of judging stability conditions is provided by the method of investigating throughflow conditions. The basis of such investigations is provided by the fact that, under identical conditions, repeated determinations of throughflow waves will yield somewhat differing results. This deviation is caused partly by measurement errors and partly by the instability of flow. Therefore, if greater accuracy is required, then averaging of the results of several measurements will yield a better solution. The deviation of the data of individual measurements from the thus determined average throughflow curve will be characteristic of the instability of flow.

4.10 Hydraulic characterization and evaluation of the settling space

The most frequently applied methods of evaluating processes taking place in settling structures are the following: a) methods based on the determination of throughflow waves and curves; b) the method of velocity-distribution measurements; c) the method of discharge-distribution measurements; and d) characterization on the basis of suspended solid distributions. Some Hungarian and other examples of the application of the above methods will be presented in this section.

Figure 4.18 shows some characteristic throughflow curves, on the basis of the work by Camp (1953), for different basin shapes. The Figure also shows the curves corre-

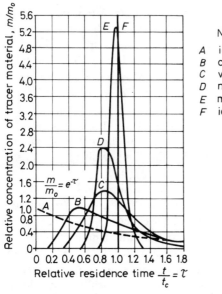

NOTATION:

A ideal tank reactor
B circular basin
C wide rectangular basin
D narrow rectangular basin
E mixed basin with baffle walls
F ideal pipe reactor

Fig. 4.18. Dimensionless throughflow waves prototype settling basins of different type (after Camp 1953)

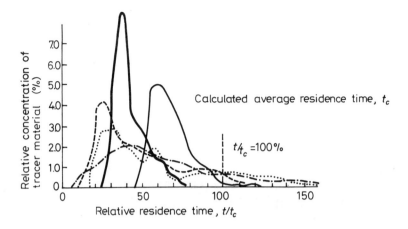

Hydraulic efficiency η = 33 %   1 ----- Circular basin
„Imhoff" tank , η = 48 %   2 —·—·— „Imhoff" tank
Longitudinal throughflow, $h:L$=1:10, η = 39 % 3 ········· Longitudinal throughflow, 1:10
Longitudinal throughflow, $h:L$ = 1:45, η =39% 4a —— Longitudinal throughflow, 1:45T
Longitudinal throughflow, $h:L$=1:45, η=61%   4b —— Longitudinal throughflow, 1:45T

Fig. 4.19. Throughflow waves prototype settling basins of different type (after Knop 1951)

sponding to ideal pipe and tank reactors, as the extreme situations, within which the throughflow waves of real settling structures represent the transitional situations. Throughflow waves measured in prototype settling structures are shown in *Fig. 4.19* indicating the respective hydraulic efficiencies ($\eta_m = 100\ t_{max}/t_c$). The relevant Hungarian literature offers several similar examples (Muszkalay and Vágás 1954).

An illustrative complex method is presented in *Fig. 4.20* Isotope tracer was used in Dortmund-type settling basins for the determination of throughflow waves. The throughflow curves corresponding to eight outflow points around the circumference of the structure provide for a reliable determination of the hydraulic conditions of the basin. The curves also indicate that the loading on the structure is asymmetrical, which is unfavourable.

Figures 4.21/a—b and *4.21/c* are presented to visualize the velocity distributions in settling basins. *Figures 4.21/a—b* correspond to the Kowal-type solution of water inflow and outflow. The purpose of the model experiments by Kowal (in Kalman 1966) was to increase the hydraulically effective settling length by finding favourable inflow and outflow arrangements. *Figure 4.21/c* shows the flow pattern in a settling tank with vertical throughflow on the basis of Sifrin's scale-model experiments. In *Fig. 4.22* the velocity distribution is characterized also numerically. The velocity distributions determined in the settling basin of the sewage treatment plant at Künsnacht,

a) Throughflow waves at the measurment points

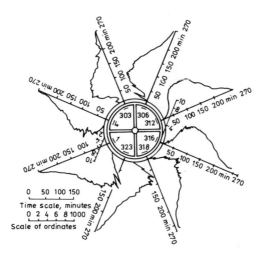

b) Distribution of tracer material

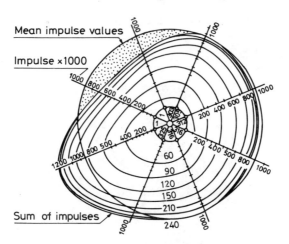

Fig. 4.20. Evaluation of throughflow investigations in Dortmund basins (after Rhode, in Pallasch and Triebel 1967)

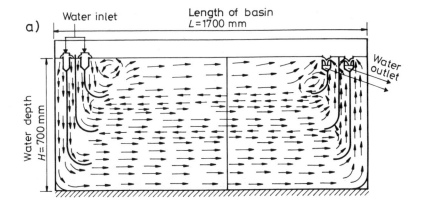

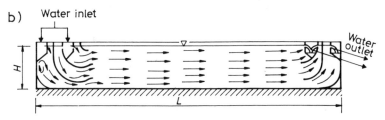

Fig. 4.21/a–b. Velocity distributions in the case of Kowal-type inlet and outlet arrangements

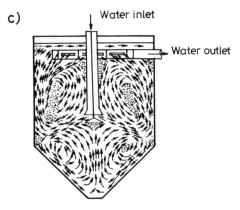

Fig. 4.21/c. Flow pattern in vertical flow basin (after Sifrin, in Kalman 1966)

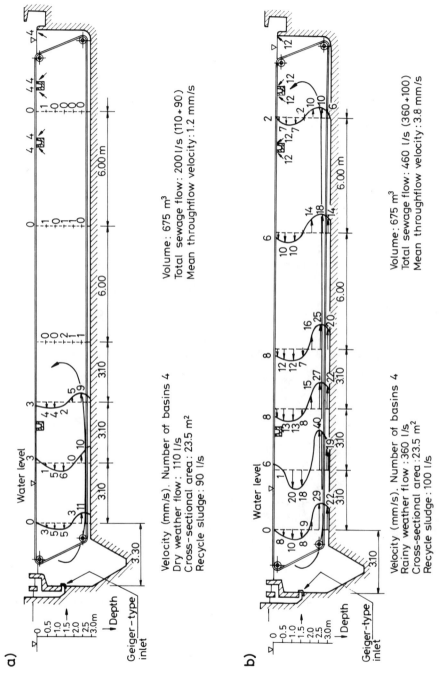

Fig. 4.22. Velocity distribution in longitudinal-flow settling basins for dry and rainy weather flows (after Hörler 1969)

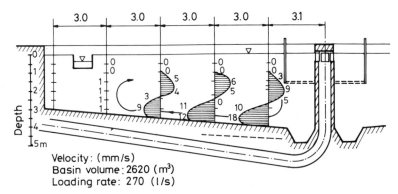

Velocity: (mm/s)
Basin volume: 2620 (m³)
Loading rate: 270 (l/s)

Fig. 4.23. Unfavourable velocity distribution in Dorr-type settling basin (after Kalman 1966)

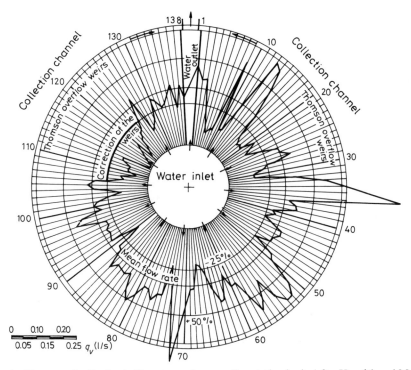

Fig. 4.24/a. Flow rate distribution in Dorr-type primary sedimentation basin (after Horváth and Muszka-lay 1969)

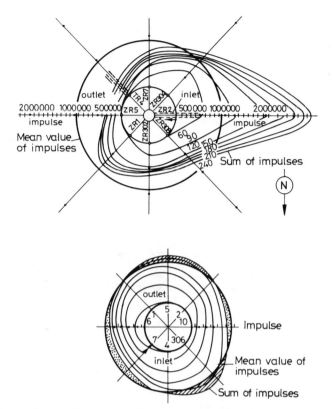

Fig. 4.24/b–c. Even and uneven flow-rate distribution in the outlet trough of a circular settling basin (after Pallasch and Triebel 1967)

near Zürich, for flows in dry and rainy periods, indicate unfavourable flow conditions (high bottom-flow velocities, backward flow in the upper zone, vortices and hydraulic jumps). Another example is shown in *Fig. 4.23* for Dorr-type settling basins, on the basis of measurements by Kalman (1966). In *Fig. 4.24/a*, the results of a Hungarian experiment with a Dorr-type settling basin, carried out by Muszkalay and the present author (1969), are shown. The circular diagram presents an example of characterizing flow conditions and discharge-proportional velocity distributions, by the distribution of flow over the weir (a series of Thomson weirs) of the inlet structure. The distribution of the $Q_v = 26.1$ l/s discharge among the 138 pc V-shaped weirs is rather irregular due, for example, to the inaccurate shape of the V weirs, to the effect of wind and floating debris, etc. The locations of the V weirs that carry 50% more or 25% less flow than the average value are also marked on the figure. Correction of the weirs at these locations is an unconditional requirement.

Results of investigations carried out with similar objectives but using radioactive tracer are shown in *Figs 4.24/b* and c, on the basis of foreign literature. *Figure 4.24/b*

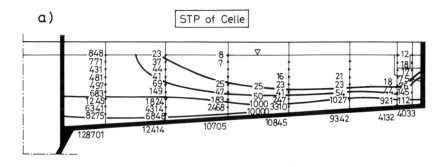

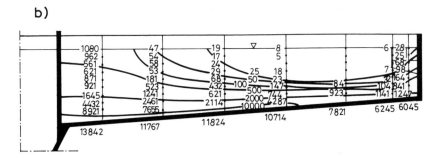

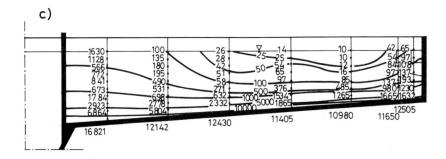

Parameter= G_c (mg/l)

Nr.	1 Q_v (m³/h)	2 t_c (h)	3 T_f (m/h)	4 R (m³/h)	5 G_c (g/l)	6 G_R (g/l)
a)	360	5.0	0.48	300	5.62	12.87
b)	450	4.0	0.60	300	5.52	12.80
c)	600	3.0	0.80	300	5.10	14.72

Fig. 4.25. Distribution of sludge concentration in a circular final settling basin (after Pflanz 1966)

indicates a strongly asymmetrical velocity distribution, while *Fig. 4.24c* shows favourable flow conditions.

Finally, an example on the basis of Pflanz's (1966) research results, is presented of the distribution of the suspended-solid content, indicating that the suspended-solid content of the sewage might induce secondary currents. *Figure 4.25* shows the distribution of suspended-solid concentrations, with isoconcentration lines, in a circular final settling tank, for three different average residence times. The stratification of settling sludge is well indicated by the figure. Lower suspended-solid concentrations in the upper zones correspond, obviously, to longer residence times. The operational parameters corresponding to the three alternatives are tabulated on the Figure. The investigations were carried out in the treatment plant at Celle (Germany). The main parameters of the structure were: $V = 1800$ m^3; $A_s = 790$ m^2; $H = 2.27$ m; $D = 33$ m; $H:R = 1:7.3$; for $t_c = 3$ h; the loading rate on the overflow weir is 5.55 m^3/h·m; $Fr = 9.02 \times 10^{-9}$.

4.11 Secondary currents

Secondary currents developing in settling structures strongly affect the hydraulic and technological efficiencies. Secondary currents may be due to the following conditions: a) differences in sewage temperatures: b) differences in the physical properties of the sewage (e.g., in suspended-solid content); c) external effects (e.g., wind effects; convective currents due to uneven warming up, etc.); d) mechanical equipment (e.g., scrapers); e) certain structural elements. Disturbing effects due to the above factors may be so excessive that they basically depreciate the operation of the structure. Some of these effects will be discussed below in more detail.

Effect of temperature differences

The temperature of sewage water discharged into a settling basin and that of the sewage water already in the basin usually differ from each other. Even a temperature difference of 1—2 °C might cause secondary currents. These currents are actually caused by density differences, thus they might be termed density currents. If, for example, a temperature difference of $15 - 12 = 3$ °C is considered, then the density difference of the water will be 0.4 ‰. In this case, the settling velocity of a "water sphere" would be 40 m/h, a significant value.

The effects of secondary currents due to temperature difference may eventually also vary with the type of the structure. To illustrate this the schemes of density currents in some types of settling basin are shown in *Fig. 4.26*. Settling basins of the longitudinal throughflow type are usually favourable from the hydraulic point of view, since the design conditions can easily be implemented in the prototype, due to the relative simplicity of such structures. It is seen from *Fig. 4.26* that dead spaces due to temperature differences do not affect the effective length of the structure, and only the throughflow cross-sectional area will be reduced. Comparing alternatives *a)* and *b)*,

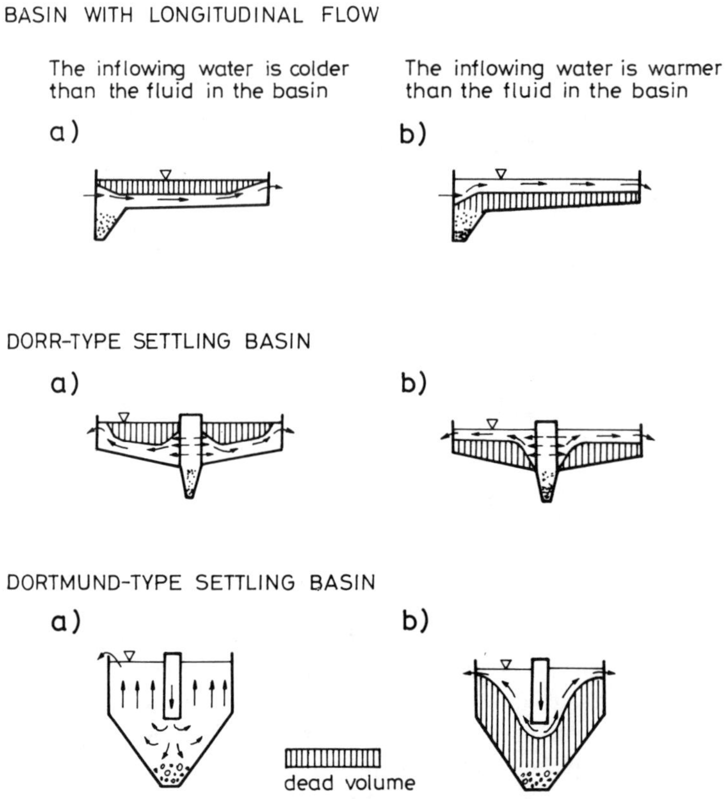

BASIN WITH LONGITUDINAL FLOW

The inflowing water is colder   The inflowing water is warmer
than the fluid in the basin   than the fluid in the basin

a)   b)

DORR–TYPE SETTLING BASIN

a)   b)

DORTMUND-TYPE SETTLING BASIN

a)   b)

dead volume

Fig. 4.26. Effects of temperature difference on the flow pattern of settling basins (after Saitenmacher 1965)

it can be stated that the arrangement shown in *Fig. 4.26a* is more favourable from the viewpoint of settling efficiency, since in this case particles settle directly onto the bottom of the basin from where they can be continuously removed. In case *b*, however, particles must travel along longer pathways and settle through the dead space, while they can more easily get into the effluent from the upper layers of the basin. In practice, however, case *a* occurs more frequently since the density of inflowing water is usually higher than that of the water in the settling basin. Similar current patterns are expected to develop also in Dorr-type facilities, as shown in *Fig. 4.26*. In this case also, the *b* version is the more unfavourable. Density currents transport settling particles into the upper layers of the basin, thus reducing the settling efficiency and offering a chance of the particles leaving the system with the outflowing water. The settling efficiency might be further reduced by wind effects, causing an uneven loading pattern.

In the case of Dortmund-type settling structures, the flow conditions will be altered considerably, due to the vertical throughflow, in comparison with the above-mentioned structures. As is indicated by *Fig. 4.26*, no dead spaces will develop in case *a* since the inflowing fluid of higher density moves towards the bottom from where it rises again, pushing the layers of lesser density in front of it along the full cross-sectional area of the structure. In case *b*, however, short-circuiting occurs, thus turning the bulk of the volume of the settling basin into a dead space. Both versions indicate basic faults. Although the hydraulic efficiency in case *a* is (theoretically) 100%, this arrangement is still unfavourable since the material settled out might be scoured up again from the bottom. This indicates the need of always reviewing the hydraulic effectiveness from the viewpoint of technology.

It is to be noted that density differences might also cause hydraulic problems in laboratory and prototype experiments when using tracers. The effects of tracer materials, other than "ideal" should be either eliminated or taken into account. The above considerations justify the need for applying radioactive tracers in hydraulic investigations.

Effects of differences in concentrations of suspended solids

The suspended solids concentrations of sewage flowing into primary settling basins might differ significantly from that already in the basins. Even higher concentration differences are encountered in final settling basins. Moreover — as was mentioned in relation to *Fig. 4.25* — the effects of differences in concentration might also be significant in the settling structure proper.

The effluent of aeration basins usually contains 4—6 g/l of solids, while the concentration of recirculated sludge is 8—15 g/l.

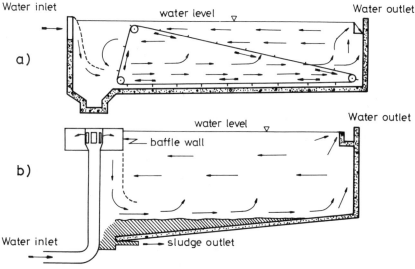

Fig. 4.27. Secondary density currents in longitudinal and radial-flow settling basins (after Anderson 1945)

In respect to final settling structures, attention to the effects of density currents was first drawn by Anderson (1945). The results of his observations are shown in *Fig. 4.27*. Following his advice, the overflow weirs of the circular settling basins of a treatment plant in Chicago were installed not at the outer circumference of the basins (in the upward flowing zone of high sludge content), but at 2/3—3/4 of the radius (in the zone of lower suspended-solid content). *Figure 4.28* illustrates the "waterfall effect". Due to the baffle wall and to density differences, inflowing water "falls" downward, scouring the deposited sludge. The favourable effects of the inlet structure arrangement proposed by Kowal (in Kalman 1966), should be referred to in this respect.

Wind effects

Settling facilities applied in sewage-treatment technologies are almost exclusively constructed in the open air and thus subjected to the effects of the weather (temperature fluctuations, wind, rainfall, etc.). In this section certain practical considerations related to the effects of wind will be presented.

Upon the effect of wind, near-surface currents might be significantly modified. Depending on the size of the structures and on the wind velocity, the increase of flow

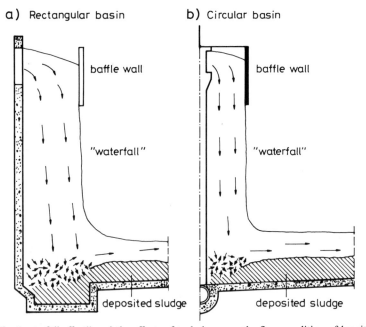

a) Rectangular basin   b) Circular basin

baffle wall

"waterfall"

deposited sludge   deposited sludge

Fig. 4.28. The "waterfall effect" and the effects of turbulence on the flow condition of longitudinal and radial-flow settling basins (after Sawyer and King 1969)

velocities might be as high as 30—40 cm/s. In estimating wind-induced surface currents in lakes, 1—3% of the wind velocity is usually taken into account. Thus, for example, a wind velocity of 10 m/s causes surface flow velocities of magnitudes of 10—30 cm/s. Increased flow velocities might cause wave motion, which might also play a certain role in settling structures.

Wind effect can be especially significant in large Dorr-type structures. Wind set up might be in the order of a few millimeters. This causes continuously changing loading rates on the overflow weirs placed along the circumference of the structure. This might result in an uneven distribution of removal efficiencies and residence times among the different parts of the basin. According to the measurement results of Gruhler (1962), wind effects might cause differences of 5—8% in the treatment efficiency, as calculated with the method of throughflow waves. It should be noted, however, that on the one hand, this difference is not too significant and on the other, the uncertainty or error in determining the throughflow wave is of the same magnitude.

In the Dortmund-type settling basins of longitudinal throughflow, wind effects are of even lesser significance since their size and design partially prevent the development of wind-induced currents. With respect to depth, the shallow settling structures are more influenced by wind effects than the deeper ones.

During the trial operations of the treatment plant at Pécs, Hungary, we made investigations on the Dorr-type settling basin of 1100 m^3 volume and 42 m diameter in order to estimate wind effects. It was found that increasing flow velocities in the direction of the wind were superimposed on currents induced by the device serving for the removal of floating debris. The modification of surface currents was easy to observe (without tracer materials), by visually following the movement of floating solids. Surface currents of 25—30 cm/s were measured, in association with wave heights of 2 cm.

It was finally concluded that the effect of wind upon the processes of settling structures is usually not too significant. However, if one wishes to eliminate wind effects, the planting of non-deciduous trees across the dominating wind direction might offer a solution. This process may be associated with rearrangement of the terrain.

Effects of mechanical equipment

In this respect, it is mainly the effects of the scrapers that should be mentioned. The usually considered design value is a 2—3 cm/s velocity of movement, so as not to scour up the already deposited substances. Some other standards allow (especially in primary settling basins) a 5—6 cm/s velocity. It is to be noted that density currents occurring at water inlets tend to transport deposited substances in a direction opposite to that of the scraper movement.

Effects of structural elements

Certain structural elements of settling basins (baffle walls, diversion plates, inlet structures, overflow troughs, etc.) might cause undesirable currents or dead spaces. Faulty and inaccurate construction might also cause problems. In this respect we refer to those stated in Section 4.7.

4.12 Critical sediment-driving force and critical flow velocity

Deposited solids might be set into motion once again if forces or the respective flow velocities exceed a certain critical value. Theories of sediment motion, as transformed to the case of settling basins, will be summarized below on the basis of the investigations by Camp (1955).

Shields' classical formula is (see *Fig. 4.29*)

$$\frac{T_c}{(\gamma_h - \gamma)d} = \Phi\left(\frac{d}{\delta}\right) = \beta \tag{4.12}$$

and for $\beta = $ const:

$$T_c = \beta(\gamma_h - \gamma)d, \tag{4.13}$$

where T_c is the hydraulic force that sets sediment into motion; d is the particle diameter; δ is the thickness of the laminar sheet; γ_h is the specific weight of the sediment and β is a coefficient for shape and resistance. According to Camp (1955),

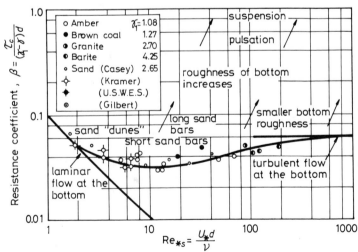

Fig. 4.29. Resistance coefficient as a function of the Reynolds number (after Shields, in Bogárdi 1972)

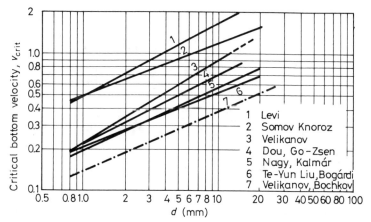

Fig. 4.29/a. Relationship between the critical bottom flow velocity and the mean particle diameter

$\beta = 0.04$ for particles of nearly identical size. Fair et al. (1968) and Eckenfelder (1970) proposed $\beta = 0.06$ for particles of different size.

The critical mean velocity v_c, corresponding to T_c, can be derived on the basis of the following considerations. On the basis of du Boys' expression (in Németh 1963) for a specific sediment-driving force $T = \gamma RI$ and considering the Darcy–Weissbach equation for friction losses, it can be written that

$$T = \gamma RI = \gamma R\lambda \frac{1}{4R} \frac{v^2}{2g} = \frac{\lambda \gamma v^2}{8g}. \tag{4.14}$$

In the critical case $T = T_c$ and $v = v_c$, i.e., by combining eqs (4.13) and (4.14) one obtains that

$$v_c = \sqrt{\frac{8\beta}{\lambda} gd \frac{\gamma_h - \gamma}{\gamma}}. \tag{4.15}$$

For example: for initial data $\beta = 0.04$; $\lambda = 0.03$ (concrete surface) and $d = 5 \times 10^{-3}$ cm, the critical flow velocity will be $v_c = 3.2$ cm/s. In the case of settling communal sewage water, velocity $v = Q_v/F = Q_v/BH$ must not exceed 1.0 cm/s. This allows a certain safety margin that will be justified by the fact that high secondary currents may be superimposed on this flow velocity in the bottom zone. Various authors have demonstrated that velocities significantly lower than 2—3 cm/s will not result in appreciably improved settling efficiency.

Figure 4.29 shows a comparison of the measurement results of various authors (after Stelczer, in Bogárdi 1972) in relating critical bottom flow velocity to the particle diameter. Kalmár, Karádi and Nagy (in Bogárdi 1972) distinguished three flow ranges in a similar approach, as functions of the thickness of the laminar sheet: the laminar range of sediment transport $d/\delta < 1$; the transitional range $1 < d/\delta < 100$ and the

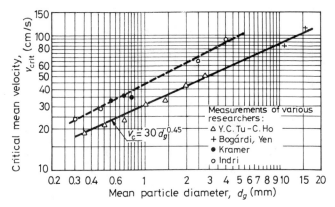

Fig. 4.29/b. Determination of critical bottom flow velocity for sand and coal particles

turbulent range, $d/\delta > 100$. In this latter range $\beta = \text{const} = 0.055$, well approximating the already mentioned $\beta = 0.06$ value.

Ingresol and coworkers demonstrated that in the case of flocculate substances (e.g., activated sludge, or sludge resulting from chemical treatment) sediment might be resuspended after settling if shear velocity $\sqrt{\tau/\varrho}$ equals or exceeds settling velocity w. Considering relationship $T = \tau = \varrho g R I$ and the Darcy–Weissbach formula, one obtains

$$w = \sqrt{\frac{\tau}{\varrho}} = \sqrt{gRI} = \sqrt{\frac{\lambda}{8}}\,v\,, \tag{4.16}$$

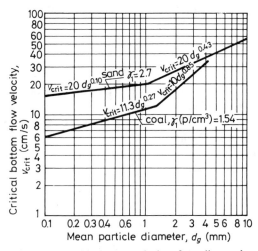

Fig. 4.29/c. Relationship between the critical mean velocity of a sedimented water course and the mean particle diameter of the sediment

and

$$v = \sqrt{\frac{\tau}{\varrho} \frac{8}{\lambda}} = w \sqrt{\frac{8}{\lambda}} \; ; \quad \frac{v}{w} = \sqrt{\frac{8}{\lambda}} = \frac{C}{\sqrt{g}} , \tag{4.17}$$

where C is the velocity coefficient of the Chezy formula, $C = \sqrt{\frac{8g}{\lambda}}$.

If, for example, $\lambda = 0.03$, then $v/w = 16.3$. In this case, the resuspension of already settled sludge can be expected at an average throughflow velocity $v = 16.3w$. As an easily applicable rule of thumb, Fair, Geyer and Okun (1968) suggested the use of the formula $v/w = 10$. For deposits of mixed composition (particulate and flocculate) the velocity ratio v/w varies between 20 and 40, according to different authors, where $v = v_{crit}$ the critical flow velocity above the bottom (that initiates resuspension).

Finally, a summary of the results of different authors is presented (after Bogárdi 1972) in Figs 4.29/a, 4.29/b and 4.29/c for estimation of the critical mean and bottom flow velocities.

4.13 Sludge index

The settleability conditions of activated sludge are usually characterized by various indices, such as the Mohlmann-, Donaldson index, etc. In Hungary, the Mohlmann index I_M is the most widely used; this is the ratio of sludge volume V_{30} (ml/l), as determined in a 1000 ml glass cylinder after 30 minutes settling time, to the concentration of the sludge G_s (g/l). Consequently, the smaller the value of sludge index I_M, the better the settling properties. For average communal sewage, the value of I_M varies between 30 ml/g and 80 ml/g. The favourable range is $I_M < 100$ ml/g. A recent Hungarian technical guide proposes 120—150 ml/g as an acceptable range.

In applying sludge indices it is implicitly assumed that the index value is constant for a given sewage, even if V_{30} and G_s vary. Nevertheless, this holds true only when the rates of change of V_{30} and G_s are identical.

More detailed investigations demonstrated that the I_M value of a given sewage remains constant only within certain limits. According to the measurements by Stobbe (1964), I_M can be considered constant for a given communal sewage if the condition $V_{30} < 200$ prevails. This also means that the value of the sludge index varies as a function of several parameters, and thus it cannot be considered an unambiguously characteristic measure of settleability.

Bond (in Fair et al. 1968) suggested, in this context, the introduction of the 24-hour settling volume concept. The long experimental time required for this determination makes the method rather cumbersome to use. On the other hand, fermentation/digestion processes might occur during the experiments, causing disturbances. Instead of this, Stobbe proposes dilution of the sample to achieve the condition $V_{30} < 200$ ml/l.

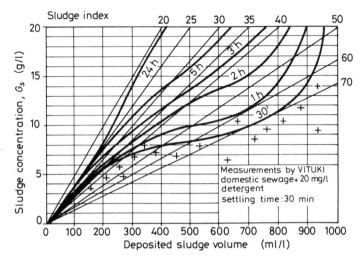

Fig. 4.30. Relationship between sludge index I_M, the dry-material content, and the settling time
(after Stobbe 1964)

In order to illustrate the above statements, *Figure 4.30* is presented, showing the relationship between solid content (g/l) and the volume of deposited sludge (ml/l), for various settling times. The Figure also includes the measurement data of an experiment carried out in the Water Resources Research Centre VITUKI (Hungary) with domestic sewage containing fatty alkyl sulfonate. The Figure also indicates that the consideration of the limiting condition $V_{30} < 200$ ml/l is justifiable. Our practical experience indicates that the relationship between the variables V_{30} and G_s is not so marked as those in the curves presented by Stobbe (1964). In the higher ranges of V_{30} values, the variance of measurement data is rather high and their correlation requires additional consideration of other affecting factors.

4.14 Efficiency of settling basins

The hydraulic efficiency of settling basins can be determined by throughflow investigations — by constructing the throughflow curves or waves. These methods are presented in the works of Danckwerts (1953), Camp (1953) and Müller-Neuhaus (1952/53), and also in those of the Hungarian researchers Muszkalay and Vágás (1954).

The technological efficiency of settling structures in continuous operation can be calculated by the relationship $\eta = 100(C_0 - C_e)/C_0$, where C_0 is the suspended solid concentration of the influent, while C_e is that of the effluent. The settling efficiency is frequently defined, for particulate matter, also by the following relationship

$$\eta_s = \frac{h}{H} = \frac{w}{w_h} = \frac{w}{T_f}. \tag{4.18}$$

To interpret this formula, let us consider first the pathway of motion of the settling particles, which is defined by the sum of the vectors of settling velocity and through-flow velocity. For $w-$const and $v=$const, the pathway is a straight line as was shown in *Fig. 4.1*. For a settling velocity of particles with 100% settling efficiency, the relationship $w \geq T_f = w_h$ corresponds. In the range of $w < T_f = w_h$ only partial settling is expectable. Particles in the range $w < w_h$ will settle out from a depth $h = wt_c$, but not from levels higher than h. Therefore, the percentage ratio of settleable particles just equals $h/H = \eta_s$. It can be also concluded from eq (4.18) that — in accordance with the Hazen concept — settling efficiency η_s is independent of depth H and average residence time t_c.

The settling efficiencies of various settling basin designs will be discussed below (after Fair, Geyer and Okun 1968).

Let us first consider settling basins of the longitudinal throughflow type, using the notations of *Fig. 4.1*:

$$\eta_s = \frac{h}{H} = \frac{w}{w_h} = \frac{wt_c}{w_h t_c} = \frac{wA_s}{Q_v}. \tag{4.19}$$

The same result is obtained on the basis of the geometrical dimensions of the Figure $(dh/dl = w\,dt/v\,dt = const;\ h/l = w/v;\ h/H = (w/v)\,(l/H) = (w/B)\,(vHB) = wA_s/Q_v = \eta_s)$.

A similar way of thinking can be followed in the case of settling basins with radial throughflow (*Fig. 4.1/c*). In this case, throughflow velocity v decreases along the radial distance, thus $v \neq const$: $v = Q/2\pi rH$ and the settling pathway of particles follows a curve.

Further, it may be written that: $dh/dr = w/v = 2\pi rHw/Q$, and $h/H = w(r_0^2 - r^2)Q = = wA_s/Q = \eta_s$. Namely, the same relationship is obtained also for the radial structures.

Rearranging eq (4.19) and making use of the above considerations:

$$A_s = \frac{Q_v \eta_s}{w} = \frac{Q_v \eta_{s\%}}{100\,w} = LB, \tag{4.20}$$

the design formula is obtained.

Summarizing: In accordance with Hazen's (1904) concept, settling efficiency η_s is independent of water depth H and average residence time t_c for both longitudinal and radial structures, when considering the settling of particulate substances. If $w \geq w_h$ the settling efficiency reaches 100%. For the case of $w \leq w_h$ the efficiency will be defined by eq (4.19). For $w \geq w_h$ the 100% settling efficiency still remains achievable by splitting the total settling depth into partial depths smaller than $h = wt_c$ (storeyed or lamellar settling facilities), thus allowing the deposition of parti-

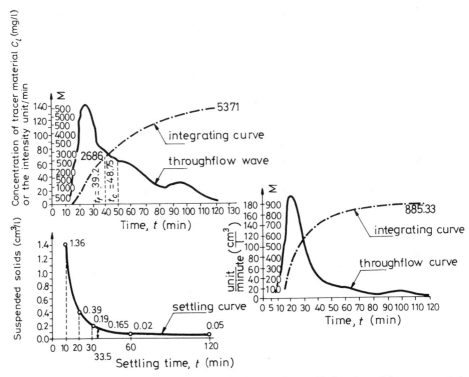

Fig. 4.31. Determination of settling efficiency by simultaneously considering throughflow wave and the settling curve

cles of settling velocities smaller than w_h. Thus, with decreasing w, the settling distance h should be also reduced.

In settling structures with vertical throughflow only those particles will settle out for which the condition $w \geq w_h = T_f$ is valid. Consequently, no settling is expectable in the $w < w_h$ range, as opposed to the case of facilities with longitudinal throughflow. In the case of vertical throughflow T_f is identical with the mean flow velocity v, i.e., both the physical meaning and the unit of T_f become the mean velocity.

Knop (1951) characterized the operation of settling basins by simultaneously considering the throughflow wave and the settling curve $C_l = f(t)$. The procedure of this author is illustrated by *Fig. 4.31*. The throughflow wave and its integrating curve are combined with the settling curve by multiplying the relevant ordinates by each other. Suspended solids not deposited are characterized by the ratio of the ordinates of the peak of the original integral curve to that of the combined curve (885.33/5371 = 0.165), and the corresponding settling time (33.5 min) is found from the settling curve. Dividing this time by the calculated residence time t_c and multiplying by 100, the settling efficiency is obtained [(100 × 33.5)/48.75 = 68.6%].

4.15 Lamellar and tubular settling facilities

Lamellar and tubular settling devices were developed on the basis of the same hydraulic principles. It is known that laminar and stable flow conditions are assumed for assuring favourable operating conditions. According to Fischerström (1955), these conditions are defined by the following Reynolds and Froude numbers:

$$\text{Re} = \frac{vR}{v} \leq 500 \qquad \text{Fr} = \frac{v^2}{gR} > 10^{-5} \qquad (4.21a\text{---}b)$$

Generally, these conditions will not hold for the conventional settling facilities. The usual ranges in settling facilities are:

$$10^3 < \text{Re} < 2.5 \times 10^4$$
$$10^{-6} < \text{Fr} < 2.6 \times 10^{-5}.$$

The above-indicated problem stems from the condition that with the increasing inflow Q and thus with increasing flow velocity v, Fr changes favourably whereas Re reacts unfavourably. Further analysis reveals that a decrease of the hydraulic radius R (i.e., an increase of the wetted circumference) acts favourably. This increase of the wetted circumference K can be achieved by installing separation walls, sets of tubes, etc. With the decrease of hydraulic radius R, a further favourable condition is a decrease of the settling depth, that is, a decrease of the zone of obstructed settling. In traditional settling basins, the suspended-solid concentrations increase considerably with the settling depth, thus creating the condition of obstructed settling.

It is to be noted that the origin of the above principle can be traced back to the classic work of Hazen (1904), and it can be found also in the later studies by Camp (1946). Practical implementation, however, was delayed until the appearance of light construction materials (mostly plastics).

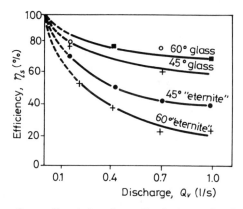

Fig. 4.32. Settling efficiency of paper fibres in lamellar settling basin as a function of the sewage discharge rate

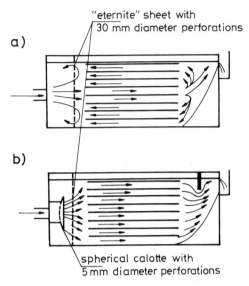

Fig. 4.33. Faulty and correct inlet arrangements in lamellar settling basin

Table 4.2. Some design and operational data for tubular settling facilities
(Heckel and Korn in Bogárdi 1972)

Type of water to be treated	Dry material content (influent) (mg/l)	Pipe diam. d (mm)	Pipe length l (mm)	l/d	Inclination angle α°	Surface loading rate T_f (m/h)	Dry-material content (effluent) (mg/l)
1	2	3	4	5	6	7	8
Surface water							
— flocculated		50	500	10	55	14	<50
($Al_2SO_4 \cdot 18\ H_2O$)	<200						
— flocculated		50	500	10	55	12—16	<50
($FeCl_3 \cdot 6\ H_2O$)	<200						
— non-flocculated	<200	100	1000	10	55	14	20—30
Groundwater	Fe: 6—12	50	500	10	55	12	Fe: 1.0—4.5
Waste water		50	500	10	55	16—18	90—100
— mechanical	<450						V_{30} 0.3—0.3 ml/l
treatment	V_{30} <10 ml/l						
— biological							
treatment							
a) activated	>2000	100	1000	10	55	7	<30
sludge	<2000	80	800	10	55	9	<30
b) trickling							
filters	<300	80	800	10	55	10—12	40—50

Remark: V_{30} = suspended solids that can be settled during 30 min (ml/l)

Today, lamellar settling basins are frequently applied (for example, in oil-traps of the Shell type). Consequently, these facilities will be mentioned also in relation to flotation facilities. In this context, the results of a Hungarian experiment will now be presented. A lamellar settling facility was tested by Szalay (1967) for the settling of waste-water containing paper fibres. The effects of discharge Q and the angle of inclination of the lamellas (glass and eternite plates) are shown in *Fig. 4.32*.

Figure 4.33 presents two alternative solutions (a good one and a bad one) for the inlet and outlet arrangements as used in these experiments.

Tubular settling devices are the most recent settling facilities. The hydraulic conditions are well described in a study by Yao (1970). Practical applications were published by Heckel and Korn (in Bogárdi 1972). *Table 4.2* is presented (after the latter authors) containing some important operational and design data for tubular settling facilities.

In tubular settling facilities the hydraulic radius can be reduced to an appropriate value, thereby also reducing the settling pathway. The proposed diameter of the tube is $d = 50$—100 mm (but it may perhaps be in the range of 25—150 mm). This assures ranges of $Re < 10$ and $Fr > 5 \times 10^{-5}$. According to Yao (1970), the slimness factor l/d should be 20 and the angle of inclination $\alpha < 40°$. Heckel and Korn (in Bogárdi 1972) proposed values of $\alpha = 50°$—$60°$ and $l/d = 10$, as being more favourable for slide-down of the deposited sludge.

Finally, it should be noted that tubular settling facilities can be applied for both new treatment plants and for extending the capacities of existing structures (sets of plastic tubes or insertion pieces made of corrugated sheet can, when appropriately installed, improve the capacities of overloaded settling structures).

5 Hydrocyclones

The basic principle of operation of hydrocyclones is that sewage is discharged tangentially into a circular basin, thus inducing circular flow around the vertical axis of the basin. Suspended solids and non-water-soluble liquids will then be separated, due to the centrifugal force, according to their densities. Substances of higher specific weight move towards the wall of the basin where they concentrate and settle out. As opposed to this, substances of lesser specific weight will accumulate in the middle parts of the basin.

Equipment operating on the basis of the hydrocyclone principle has recently found increasingly widespread use in sewage-treatment technology such as grit chambers, grease, tar and oil traps, etc. They are especially suitable for removing waste metals and minerals from industrial waste waters. Though less frequently, they are also used as thickeners or separators.

There are several different designs of hydrocyclone, but they can all be grouped into two basic categories:

a) closed cyclones with inflow under pressure, and
b) open hydrocyclones where gravity determines the water motion.

5.1 Hydraulic principles

The flow pattern of hydrocyclones is basically determined by two different types of water motion. A vortex motion that results in circular flow pathways, in a horizontal cross-section of the basin. This is considered as the "primary flow" of the equipment and plays a deterministic role in the operation and in the treatment efficiency. There is a "secondary flow" as well, between the inlet and outlet devices of the structure, that can be characterized by the average residence time $t_s = V/Q_v$. *Figure 5.1* shows the variation of tangential flow velocities together with that of the secondary currents. The flow resulting from the superimposed primary and secondary currents depends on the basin geometry and on the arrangement of inlet and outlet pipes and follows a helical pathway.

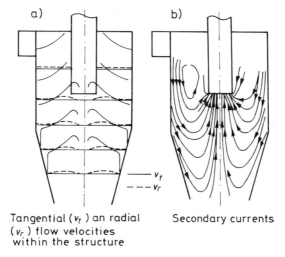

a)

b)

Tangential (v_t) an radial (v_r) flow velocities within the structure

Secondary currents

Fig. 5.1. Flow patterns in hydrocyclones (after Pattantyus 1958)

The flow conditions can be discussed on the basis of the same principles as in the case of aerated sand traps. Suspended solids in the waste water become separated and broken up into smaller pieces, due to the mechanical effects of the flow. The basic design task is to determine the diameter, d_{min}, of the smallest settleable particles. The value of d_{min} depends on many factors and the following theoretical considerations can be taken into account in deriving a simple relationship.

Let us consider a suspended particle on which two forces act; the centrifugal force:

$$C = ma = \frac{d^3 \pi \gamma v_t^2}{6gr},$$

(5.1)

where v_t^2/r is the centrifugal acceleration, and the friction force:

$$S = 3\pi\eta dw_r.$$

(5.2)

In eqs (5.1) and (5.2) r is the radius of the circular flow pathway of the particle (m); v_t is the tangential component of the flow velocity (m/h); w_r is the radial component of the velocity difference between the particle and the water (m/h); and d is the diameter of the particle (m).

In eq (5.2) the validity of the Stokes theorem has been assumed.

In a hydrocyclone, theoretically, those particles can be settled out for which the resultant radial force points towards the wall of the basin. In a boundary situation, the above two forces balance each other and this condition defines the size d_{min} of the smallest settleable particle, that is

$$\frac{d_{min}^3 \pi \gamma v_t^2}{6gr} = 3\pi\eta d_{min} v_r,$$

(5.3)

whence

$$d_{\min} = \sqrt{18\frac{\eta g v_r}{\gamma v_t^2}\, r}\,.$$ (5.4)

The above formula expresses the following conditions:

— the higher the specific weight of suspended particles;
— the smaller the viscosity of the flowing medium;
— the smaller the diameter of the hydrocyclone [i.e., in eq (5.1) centrifugal force C increases with the decrease of basin radius r];
— the smaller the radial velocity v_r (that also depends on residence time $t_s = V/Q$, i.e., on the geometry of the basin and on the hydraulic load);
— the higher the tangential velocity v_t (that depends, for example on the pressure and velocity of the inflowing water and thus on the size of the inlet pipe) and
— the smaller the diameter of the smallest settleable particle.

It is easily conceivable that eq (5.4) can also be written in terms of the Froude and Reynolds numbers:

$$d_{\min} = r\sqrt{18\frac{\mathrm{Fr}_r}{\mathrm{Fr}_t\,\mathrm{Re}_r}}\,,$$ (5.5)

where

$$\mathrm{Fr}_r = \frac{v_r^2}{gr}; \qquad \mathrm{Fr}_t = \frac{v_t^2}{gr}; \qquad \mathrm{Re}_r = \frac{v_r r}{v}\,.$$ (5.6a—c)

Thus, the process of suspended-solid separation of a hydrocyclone is a function of the above dimensionless variables.

Tangential velocity has a specially important role, indicated by the fact that among the terms under the square-root sign only v_t has an exponent that exceeds unity. It is to be noted that in the above equation v_r replaces w_r because in the case of $C = S$ the radial displacement of a particle is zero, and thus the relative velocity will be equal to the radial velocity component v_r of the medium in motion. The radial velocity component of the particle w_r can also be expressed from the above balance equation, and can be considered as the "settling", i.e., the separation velocity:

$$w_r = \frac{d^2}{18\eta}(\varrho_1 - \varrho)\frac{v_t^2}{r}\,.$$ (5.7)

Substituting the design values of v_r, v_t and r into the above equation, the value of the separation velocity can be obtained. More detailed analysis of the velocity distribution provides further information on the initial condition.

Similarly to the analysis of sand traps with air injection, the basic relationship can be that of the circulating flow around a vortex streamline, in which the linear relationship

$$v_t = k_1 r$$ (5.8)

describes the flow velocity, while in its vicinity

$$v_t = \frac{k_2}{r},$$ (5.9)

a hyperbolic relationship prevails. The simultaneous approximate formula that describes both conditions is obtained as

$$\frac{v_t}{v_{max}} = \frac{2}{\dfrac{r}{r_m} + \dfrac{r_m}{r}}$$ (5.10)

in which the dimensionless relationship ratio v_t/v_{max} depends — in principle — exclusively on the ratio r/r_m.

In the above relationship, v_{max} is the maximum value of tangential velocity v_t (m/h); and r_m is the radius corresponding to v_{max} (m).

Equation (5.10) offers a possibility of determining the distribution of tangential velocities. It also allows the formulation of the relationship $r = f(v_t)$:

$$\frac{r}{r_m} = \frac{v_{max}}{v_t} \pm \sqrt{\frac{v_m^2}{v_t^2} - 1}.$$ (5.11)

Combining eq (5.11) with eq (5.4), the variable r can be eliminated, when calculating d_{min}. Another method can also be used for this purpose which is frequently applied in studies carried out with gaseous medium. In this approach eq (5.9) is modified as

$$v_t = \frac{const}{r^n}$$ (5.12)

which can be written also in the dimensionless form:

$$\frac{v_t}{v_{max}} = \left(\frac{r_{max}}{r}\right)^n.$$ (5.13)

The value of exponent n can be determined experimentally. The results of experiments carried out with gaseous media and water yielded $n = 0.5$—0.6.

Figure 5.2 shows the results of hydrocyclone measurements (firm Escher Wyss) carried out by Weiss and Siewert (1965) with hydrocyclones used in the pulp and paper industry. Version *a*) of *Fig. 5.2* illustrates the pressure and tangential-velocity distributions for conical angles of 6° and 12° ($Q_v = 2000$ l/min; Δp pressure gradient = 17 m water column). It can be seen that the value of v_t increases towards the axis, then drops to zero at the close vicinity of the axis. The static head decreases and also exhibits negative values. There are no essential differences between the two versions. Version *b*) of *Fig. 5.2* shows the effect of discharge. Increasing the value of Q_v from 1200 l/min to 2600 l/min, the tangential velocities increase from 8 m/s to 17 m/s at the wall. The energy demand was trebled at the same time. It is to be noted

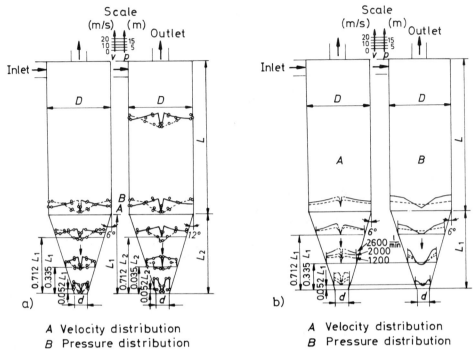

Fig. 5.2. Velocity and pressure distribution in a hydrocyclone

that the relationship between velocity and pressure was determined using the formula

$$v_t = \sqrt{2g\,\Delta p/\gamma}\,,\qquad\qquad(5.14)$$

in constructing the figures; it yielded a good approximation.

In practice, several units can be combined in a series connection, thereby forming the so-called multicyclones. Closed and open hydrocyclones operate essentially on the same principles, thus they were not distinguished in the above description.

In certain situations, however, inlet and outlet arrangements might significantly affect the flow conditions, causing constructional differences between closed and open systems.

5.2 Applications in sewage-treatment technology

The main fields of usage were already mentioned in the Introduction. In addition to this, the main principle of practical application is that hydrocyclones can be used economically only in the case of low hydraulic loading rates. In this section, the design principles and technological applications will be briefly discussed.

Figure 5.3 shows a section of a closed hydrocyclone with its most important elements. This version consists of the following main parts: a) the upper cylindrical part with the tangential inlet device; b) the conical part; c) an outlet pipe joint, and d) the outlet pipe serving for the removal of settled substances.

In this device, the influent arrives into the hydrocyclone under pressure. Substances of higher specific weight concentrate in the vicinity of the vertical and conical walls and move towards the lower part of the structure. Substances of lesser specific weight leave the system through the outlet pipe. This version is useful, primarily, as a sand trap or grit-washer (especially for removing organic substances from the deposits of sand traps operating with low effectiveness). Moreover, it can also be applied for washing the deposited sand of primary settling basins. Mineral substances (e.g., sand, gravel) separated by the hydrocyclone can usually be directly disposed off in landfills or transferred to drying beds. The effluent can be conveyed to the next element of the treatment line (e.g., to the sand trap or primary settling basin).

The efficiency of hydrocyclones should be tested experimentally in laboratory or pilot-scale experiments. Theoretical relationships are generally not sufficient for solving design tasks, although they might provide useful information for understanding the processes and for processing the measurement data. In order to illustrate the order of magnitude of characteristic data, the results of a measurement series — carried out in the former Soviet Union — will be presented below. The data relate to the treatment of the waste waters of a pipe-manufacturing plant.

The suspended-solids content of the waste water varied between 100 mg/l and 600 mg/l; 70% of the particles were in the size range of 100—500 micron. The main dimensions of the hydrocyclone applied were as follows: diameter = 375 mm; conical

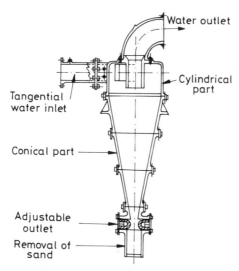

Fig. 5.3. Schematic cross-section of a hydrocyclone

angle of the lower part 10.2°; diameter of the inlet pipe $= 40$ mm; diameter of the outlet opening $= 10$ mm; influent discharge $= 45$—100 m^3/h; pressure at the inlet side 0.5—2.3 atm. Treatment efficiencies varied between 42% and 92%. The 92% efficiency was attained at a hydraulic load $Q=45$ m^3/h, with 14% of the particles being less than 100 microns in size.

In sewage treatment technology, the economic operation of closed hydrocyclones is assured when their diameter is in the range of 300—1200 mm, and the conical angle is 10°—20°. Applicable pressures vary between 0.5 atm and 3.0 atm, for larger and smaller diameter devices, respectively. The larger units are used individually, while smaller ones are connected to each other in series, and called multicyclones. The minimum settleable diameter d_{min} may be taken as 100 microns (whereas the same value for traditional grit chambers is 200 micron).

Open hydrocyclones can be constructed with diameters of several metres. In this case, however, the centrifugal force will only allow of lower efficiency of removal of suspended solids. Vertical, gravitational settling will dominate in these devices. The head loss of open hydrocyclones is about 0.5 m water column. The velocity of inflowing water is 2—5 m/s.

In the case of practical applications, the favourable properties of hydrocyclones can be summarized as follows: the removal efficiency can be higher than in the case of traditional sand traps; the smallest settleable particle diameter is smaller; the degree of separation of materials with different specific weight (settleable and floating substances) is higher; the fraction of organic substances in the deposited sand is lower; operation and maintenance costs are lower; the space requirement is (about 20 times) smaller; the device is simple and does not involve mechanical equipment (for example scrapers).

6 Grease, oil and petrol traps (flotation facilities)

Flotation facilities are mainly used, in sewage-treatment technology, for removing grease, oil and petrol from waste waters. Fields of application are: treatment of waste waters of slaughter houses, meat-processing and refrigerator plants, oil and margarine manufacturing plants, catering enterprises, crude-oil processing plants, car-washing facilities, etc.

The basic principle of the operation of flotation equipment relies on the laws of settling under gravitational conditions.

6.1 Buoyant velocities of spherical particles

The light-weight phase of the sewage in grease, oil and petrol separating equipment can be considered to contain spherical particles. In general, the size range of these particles allows the application of Stokes' law. In this case, however, the term $(\varrho_1 - \varrho)$ in eq (2.24) should be replaced by $(\varrho - \varrho_1)$. If the original form of the Stokes equation is considered, then velocities are of negative sign, that is buoyant velocities are encountered. The principles of Stokes' law have been discussed in Section 2.5.1, and will be considered as known.

Considering oil droplets of $d=0.045$ cm diameter as the critical ones, the Stokes equation can be rewritten, for practical purposes, in the following form:

$$w=0.0123 \frac{\varrho - \varrho_1}{\eta} \quad \text{(cm/s)}, \tag{6.1}$$

where

ϱ and ϱ_1 are the densities of water and oil, respectively (g/cm^3);

η is the viscosity (g/cm · s);

$0.0123 = gd^2/18$, in which g is the acceleration of gravity (cm/s^2).

Figures 6.1/a and *6.1/b* present graphical relationships between particle diameter (droplet diameter), the buoyant velocity, the density ϱ_1 and the temperature. *Figure 6.1/a* corresponds to a water temperature of 15 °C, while *Fig. 6.1/b* allows the conversion of the buoyant velocity of particles of 0.85 g/cm^3 density for water

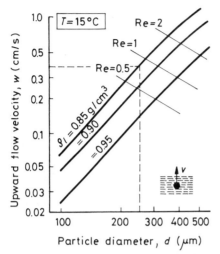

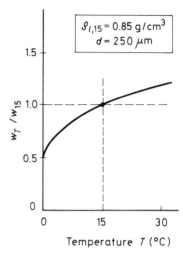

Fig. 6.1/a. Relationship between upward flow velocity and the droplet size according to Stokes (after Rumpf 1972)

Fig. 6.1/b. Effect of temperature on the upward flow velocity

temperatures other than 15 °C, in the range of $0 < T$ (°C) < 30. It is indicated that buoyant velocities decrease rapidly with the decrease of the water temperature. This phenomenon is due, mainly, to increased viscosities and — to a lesser extent — to altered densities.

It should be noted that some authors propose the calculation of buoyant velocities using Oseen's modified Stokes equation, which can be justifiable in the range of $Re \leq 5$. *Figure 6.2* compares the results of calculations with Stokes' and with Oseen's equations, respectively.

6.2 Corrections of the Stokes and Newton relationships

Similarly to the case of calculating settling velocities, calculations might be started with the Stokes equation (with appropriate corrections), even if the validity of the Stokes law is not fully guaranteed.

In calculating the velocities of settling or rising of liquid drops, the effects of the particles being non-solid can be taken into account. Namely, a certain internal circulatory motion is induced in the liquid drops, during settling or rising, which might increase the velocities. The actual buoyant velocity can be calculated with Hadamard's correction coefficient as follows:

$$w_{actual} = w_{St} \, Ha = w_{St} \frac{3\eta_1 + 3\eta}{3\eta_1 + 2\eta} = w_{St} \frac{\eta_1/\eta + 1}{\eta_1/\eta + 2/3} . \tag{6.2}$$

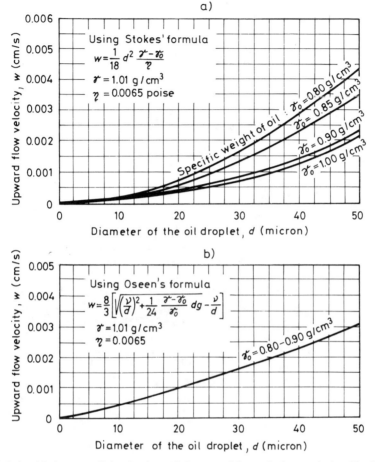

Fig. 6.2. Relationship between oil droplet size and the upward flow velocity as calculated by Stokes' and Oseen's formulae, respectively

Therefore, if $Ha > 1$, $w_{actual} > w_{St}$. For example, $Ha = 1.29$, 1.20, or 1.13 if $\eta_1/\eta = 0.5$, 1.0, or 2.0, respectively.

For oil traps, Karelin (1955) proposes the following modified relationship:

$$w = \frac{\beta g d^2}{\varphi \, 18\eta} (\varrho - \varrho_1),$$

(6.3)

where

$$\beta = \frac{4 \times 10^4 + 0.8 K^2}{4 \times 10^4 + K^2},$$

(6.4)

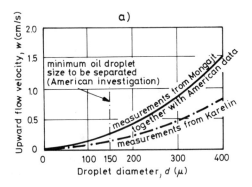

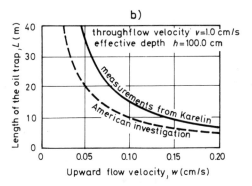

Fig. 6.3. Upward flow velocities of oil droplets, comparing American and Soviet measurement results

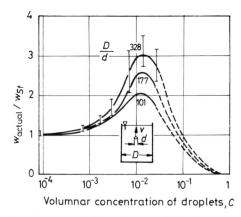

Fig. 6.4. Relationship between upward flow velocity and the volumnar concentration of droplets

K is the concentration of suspended contaminants (mg/l), and φ is a coefficient accounting for the flow conditions of the basin.

In considering longitudinal-flow oil traps, the following characteristic average parameter values can be used: $\beta=0.8$, $\varphi=1.5$, $\varrho=1.01$ g/cm^3 and $\varrho_1=0.9$ g/cm^3. *Figure 6.3* presents the combined results of Soviet (Karelin and Mongajt) and American authors.

According to Smirnov (in Mongajt and Rodziller 1958) it is justifiable to distinguish between the cases of laminar and turbulent motion, by considering Stokes' and Newton's laws, respectively, as initial conditions for calculating the buoyant velocities of liquid drops:

$$w_{\text{actual}} = \frac{d^2(\gamma-\gamma_1)}{18\eta}\left[\frac{0.9\gamma^{0.04}}{(\gamma-\eta_1)^{0.2}}\right]; \tag{6.5}$$

$$w_{\text{actual}} = 1.74\,\frac{gd(\gamma-\gamma_1)}{\gamma}\left[\frac{0.75\gamma^{0.02}}{(\gamma-\eta_1)^{0.4}}\right]. \tag{6.6}$$

There are other relationships available for calculating the corrected velocities in the transitional ranges.

Figure 6.4 is presented (after measurements by Koglin, in Rumpf 1972) to show the effects of the volumnar concentration of particles C and the D/d ratio on the variation of the ratio $w_{\text{actual}}/w_{\text{St}}$; increasing C values might substantially increase it. The 95% probability confidence limits are also shown for $D/d=328$. Rumpf (1972) attempted to explain the increasing buoyant velocities by the generation of particle complexes. The size and thus the velocity of rising of coupled or unified drops will be increased. Hydrodynamic and coagulation effects play a decisive role in this process.

6.3 Empirical relationships

Among the experimentally derived relationships, the formula elaborated by Mongajt (Vodged Institute) for calculating buoyant velocities should be first mentioned:

$$w = \alpha\,10^{0.0143\,d}\,\eta(112-93\varrho_1)\quad\text{(cm/s)}, \tag{6.7}$$

where

$$\alpha = 0.875 + 0.015\,\frac{k_a}{k_m} \tag{6.8}$$

k_a is the concentration of oil in the sewage;
k_m is the total concentration of suspended solids in the sewage.

The validity range of eq (6.7) is $10 < d(\mu) < 100$. Parameter k_m allows, to certain extent, consideration of the fact that solid contaminants present in the sewage will also affect buoyant velocities. Oil and grease particles are adsorbed by solid particles thus altering settling and rising velocities. With respect to Mongajt's measurements, *Fig. 6.3* shall be referred to once again.

6.4 Splitting up of drops

The risk of splitting up or disintegration of the droplets of the lightweight phase of the sewage is a factor that might disturb the operation of flotation. Deformation and disintegration of droplets might be due mainly to the following effects: a) shear forces; b) pressure gradients; c) diffluence at the water surface or at the surfaces of the structure. The smaller a liquid drop is the more resistant it is to deformating forces. The resistance force is proportional to the surface tension of the drop.

The smallest drop diameter d_{min} that still allows of disintegration due to the above forces can be defined. *Figure 6.5* presents relationships corresponding to the most simple basic cases: a) constant velocity gradient; b) water inlet arrangement of the Stengel-head type; c) separation at flat surfaces. The findings of this author permit the conclusion that the splitting up of drops is mainly to be expected in the vicinity of the inlet device. Consequently, it is of special importance to ensure a hydraulically favourable inlet-structure design.

a) Effect of shear force (after Karam and Bellinger)

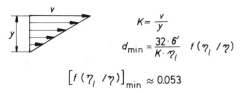

$$K = \frac{v}{y}$$

$$d_{min} = \frac{32 \cdot \delta}{K \cdot \eta_l} \; f(\eta_l / \eta)$$

$$[f(\eta_l / \eta)]_{min} \approx 0.053$$

b) Effect of pressure gradient :

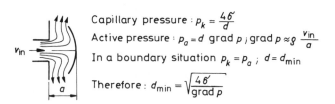

Capillary pressure : $p_k = \frac{4\delta}{d}$

Active pressure : $p_a = d \; \text{grad} \; p$; $\text{grad} \; p \approx \varrho \; \frac{v_{in}}{a}$

In a boundary situation $p_k = p_a$; $d = d_{min}$

Therefore : $d_{min} = \sqrt{\frac{4\delta}{\text{grad} \; p}}$

c) Separation of a surface :

If $a \to 0$, and the value of grad p is high then d_{min} is very small

Fig. 6.5. Splitting up of liquid droplets (after Rumpf 1972)

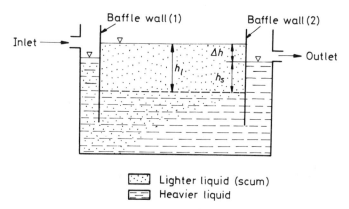

Lighter liquid (scum)
Heavier liquid

Fig. 6.6. Scheme of a flotation unit

6.5 Calculation of scum thickness

One of the important steps in designing flotation structures is the calculation of the thickness of scum.

Considering the notations of *Fig. 6.6*, the thickness of the scum, or that of the upper lightweight phase, can be estimated by making use of well-known hydrostatic relationships. The hydrostatic head at the lower plane of the floating (scum) layer is

$$p = h_1 \gamma_1 = h_s \gamma, \quad \text{or} \quad h_1 \varrho_1 = h_s \varrho \tag{6.9a—b}$$

and the head difference Δh is

$$\Delta h = h_1 - h_s = h_1 \left(1 - \frac{\varrho_1}{\varrho}\right) = h_1 \frac{\varrho - \varrho_1}{\varrho}, \tag{6.10}$$

while layer depths h_1 and h_s can be expressed as

$$h_1 = \Delta h \frac{\varrho}{\varrho - \varrho_1} \quad \text{and} \quad h_s = \Delta h \frac{\varrho_1}{\varrho - \varrho_1}. \tag{6.11a—b}$$

6.6 Hydraulic design using Hazen's concept

The principle of the hydraulic design of flotation facilities is that the through-flowing sewage should be retained for a time such as to allow the rising of suspended particles to the surface. This required retention (residence) time t_s is calculated from the condition that it should be equal to or greater than the rising time t_u of the smallest (critical) particle to be separated:

$$t_u = \frac{H}{w}, \tag{6.12}$$

where w is the upward flow velocity (rising velocity) of the particle, and H is the effective depth of the flotation structure (e.g., the rising height).

For the boundary condition $t_s = t_u$, eq (6.12) takes the form

$$t_s = \frac{V}{Q_v} = \frac{HLB}{Q_v} = \frac{H}{w} = t_u, \tag{6.13}$$

from where

$$w = \frac{Q_v}{LB} = \frac{Q_v}{A_e} = T_f. \tag{6.14}$$

The above relationship expresses Hazen's concept on the settling of particulate matter, according to which — in an ideal case — upward rising velocity w equals surface loading rate T_f. The effective area $A_e = LB$ of the flotation basin can be then designed as

$$A_e = \frac{Q_v}{w}; \qquad A_e(\mathrm{m^2}) = 0.069\,\frac{Q_v\,(\mathrm{m^3/day})}{w\,(\mathrm{cm/min})}, \tag{6.15a—b}$$

or for a given basin width B, the length of the structure can be calculated as

$$L = \frac{Q_v}{wB} = \frac{v_k H}{w}, \tag{6.16}$$

since $Q_v = Av_k = BHv_k$.

In this concept the basic design data are the sewage discharge rate Q_v and the upward rising (buoyant) velocity w. Thus, the depth of the structure does not play any role.

In designing the structure, first the smallest particle diameter to be separated should be defined, followed by the calculation of the rising velocity of this latter. In Hazen's concept the time of rising, in the boundary case, should be equal to the average retention time. According to practical experience, the average velocity of throughflow should, in the case of oil traps, meet the following requirements: $v_k < 15w$; $v_k < 1.0$ m/min. The recommendable depth to width ratio of longitudinal-flow basins is $H/B \leq 0.3$. Further design guideline data are: $v_k = 0.6$ m/min; $1.0 < H$ (m) < 2.5; $2.5 < B$ (m) < 6.0.

Finally, the pioneering measurements by Zunker (1938) for the design conditions of grease and oil traps should also be mentioned. Upward flow velocities of linseed oil particles of $w = 0.4$ cm/s were measured at a water temperature of 15 °C. The respective efficiency of removal was 95%. Zunker's investigations can be considered as classic in the field of the design of flotation facilities. He designed a grease trap on the basis of hydraulic and flow-pattern investigations in glass scale models. This design has been in use ever since, and it matches the design of casette-type settling basins of the present day. The finding that grease and oil traps can be designed on the basis of surface loading rates and the Hazen concept, belongs also among his achievements.

6.7 Design on the basis of capacity curves

The design procedure using the so-called capacity curve will be described on the basis of a study by Rumpf (1972). Let us consider, as the first version, the scheme of a longitudinal-flow flotation basin shown in *Fig. 6.7*. Assume that $w=$const and $v=$const, both in space and in time.

The equation describing the pathway of a particle is

$$\frac{dy}{dx} = \frac{w}{v}, \quad \text{whence}$$

$$y = \frac{w}{v}x + \text{const}. \tag{6.17}$$

All particles will be separated in the domain of $d>d_1$ and $w>w_1$. In the boundary case of $d=d_1$, a droplet starting from an initial point $y=0$ and $x=0$ will end up at point $y=H$ and $x=L$ (where d_1 is the critical droplet diameter).

Consider the efficiency of particle removal with the following ratio:

$$\Phi = \frac{\text{number of separated particles of diameter } d}{\text{number of particles of diameter } d \text{ in the inflow}}. \tag{6.18}$$

A still separable (flotatable) particle, entering at location $y=y_G$, will emerge at location $y=h$ on the surface. Obviously, particles entering at locations $y>y_G$ will also be separated. Then on the basis of eq (6.18) one obtains:

$$\Phi(w) = \frac{H-y_G}{H} = \frac{w}{w_1}; \quad w \leq w_1, \tag{6.19}$$

for which a uniform droplet distribution and transportation rate is assumed at the inlet. If $w \geq w_1$ then $\Phi = 1$.

In eq (6.17) const$=0$ for particles of diameter d_1 and

$$w_1 = v \frac{H}{L}. \tag{6.20}$$

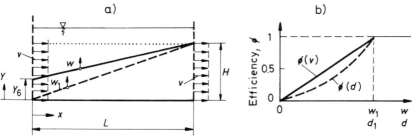

Fig. 6.7. Scheme to illustrate the efficiency curve concept, in the case of uniform velocity distribution (after Rumpf 1972)

Further, in the Stokes range w is proportional to d^2 and thus

$$\Phi(d) = \frac{w}{w_1} = \left(\frac{d}{d_1}\right)^2; \qquad d \leq d_1 . \tag{6.21}$$

The capacity curves are defined as functions $\Phi(w)$ and $\Phi(d)$, as specified by eqs (6.19) and (6.20). Plotting these equations, one obtains the capacity curves shown in *Fig. 6.7b*. For example, if $w/w_1 = 1/2$ then $\Phi = 0.5$ or 50%.

Consider, as a second alternative, the case when a vertical velocity distribution $v(y)$ is encountered which is, however, constant for all verticals of the basin. The pathway of an upward rising droplet can, in this case, be described as $dy/dx = w/v(y)$, from where

$$\int_{y_G}^{H} v(y) \, dy = \int_{0}^{L} w \, dx = wL . \tag{6.22}$$

Efficiency $\Phi(w)$ will be modified, in comparison with eq. (6.19), since the mass fluxes at different heights H differ from each other, as defined by $v(y)$. Taking this variation into consideration,

$$\Phi(v) = \frac{\int_{y_G}^{H} v(y) \, dy}{\int_{0}^{H} v(y) \, dy} . \tag{6.23}$$

Considering an average throughflow velocity v_k defined as

$$v_k = \frac{1}{H} \int_{0}^{H} v(y) \, dy , \tag{6.24}$$

and combining eqs (6.22), (6.23) and (6.24) we obtain that

$$\Phi = \frac{wL}{Hw_k} = \frac{w}{w_1}; \qquad w \leq w_1 . \tag{6.25}$$

If $w > w_1$ then $\Phi = 1$.

Comparing eqs (6.19) and (6.25), it can be concluded that efficiency is given as $\Phi = w/w_1$ regardless of the velocity distribution assumed. This means, at the same time, that velocity distribution $v(y)$ does not affect the value of w_1. In the limiting case of $d = d_1$ and substituting $\Phi = 1$ and $Q = HB v_k$, it is obtained that

$$\Phi = \frac{w_1 L}{w_k H} = \frac{w_1 LB}{Q} = 1 , \tag{6.26}$$

whence

$$A_f = LB = \frac{Q}{w_1}; \qquad w_1 = \frac{Q}{A_f} = T_f . \tag{6.27}$$

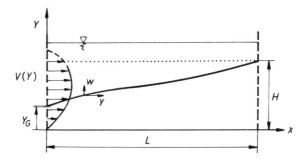

Fig. 6.8. Scheme to illustrate the efficiency-curve concept in the case of arbitrarily chosen velocity distribution

Equations (6.27) are the well-known basic design equations of the classic Hazen concept. This is one more justification of the fact that, theoretically, water depth does not play a role in the operation of such structures [i.e H is eliminated from eq (6.26) when substituting $Q = HB v_k$]. This is relatively easily conceivable when considering that for a given Q, v, increases with the decreasing H, thus decreasing the average retention time t_s. Essentially, the ratio of the time required for longitudinal travel of a particle to that required for its rising to the surface remains constant irrespective of the water depth. In practice, however, this statement holds only within certain limits *(Fig. 6.8)*.

Let us consider, as the third alternative, the case when the velocity distribution changes along the longitudinal axis of the basin, but remains constant in the lateral direction. It can be proven that in this case the efficiency curve will not be independent of the velocity distribution. The cases shown in *Fig. 6.9* will be analysed as examples.

In case a) the scum also flows, while in case b) it forms a static or stagnant cover layer. The initial uniform distribution of flow velocity changes, after distance 1, into permanent laminar or turbulent motion. The four efficiency curves corresponding to the thus defined four stretches are presented in *Fig. 6.9c*, without giving the details. It is seen that for basin type *a*, the turbulent velocity distribution yields the highest efficiency over the entire range investigated. In alternative b) this statement holds only in the range $d/d_1 > 0.7$.

The turbulent velocity distribution will not, in itself, alter the efficiency of flotation facilities unfavourably, although it might have other distrubing effects (such as the splitting up of particles).

Limit size d_1, corresponding to $\Phi = 1.0$, was the same in each of the cases investigated. Finally, it should be emphasized that practical situations might be significantly different from the alternative cases discussed above, and the latter were presented mainly to enhance the analysis and understanding of the main principles.

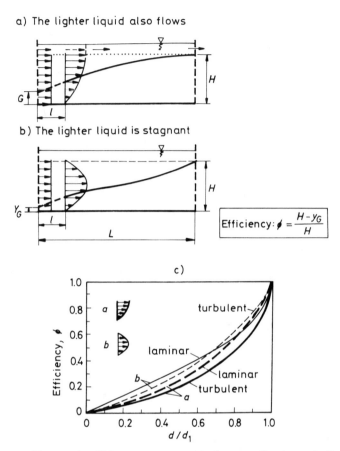

Fig. 6.9. Scheme to illustrate the efficiency-curve concept in the cases of various velocity distributions (after Rumpf 1972)

6.8 Hydraulic principles of lamellar oil traps

During the development of various settling-basin constructions, the idea of splitting up the basin depth has arisen many times. Based mainly on the principle of multistoried settling basins, lammellar flotation basins can be developed by dividing the basin volume with parallel plates set at a certain appropriate angle. The first step is to calculate the necessary settling/flotation area [eq. (6.15a)] A_s for given Q_v and w values. If the basin space is divided into vertical segments by n parallel plates, then the available basin area is multiplied by n, or — theoretically — the required area A_s can be obtained with an n-times smaller basin length or basin width. Essentially, the lamellar construction can reduce the required basin area substantially, while preserving the efficiency, and each of the bands between the plates can be considered small flotation structures of shallow depths. Thus, the settling/flotation depth will be the vertical distance between two plates. This design has the additional advantage of easily ensuring laminar flow.

Taking the above basic principles into consideration, the hydraulic justification can be given as follows. It is known that in a settling basin it is desirable to maintain laminar flow conditions which are represented by the smallest possible Reynolds number — the ratio of inertia to friction forces. At the same time, the Froude number, the ratio of inertia to gravity forces, characterizes the stability of the flow. At high Fr numbers, inertia forces increase in comparison with the force of gravity, thus increasing the stability of the flow. This can be easily conceived when considering density currents where the main force is the force of gravity. If the role of gravity is decreased in a relative sense, that of the secondary currents (such as the currents caused by density differences) decreases as well.

Then the problem of how to decrease Re numbers simultaneously with the increase of Fr numbers arises. The answer to this question leads one to the basic idea of constructing lamellar structures. Let us derive Re and Fr numbers using hydraulic radius R, as a characteristic length dimension of the basin cross-section:

$$\text{Re} = \frac{vR}{v}; \qquad \text{Fr} = \frac{v^2}{gR}. \qquad (6.28\text{a—b})$$

For a given basin, $R = A/C_w$, where A is the wetted cross-sectional area and C_w is the wetted circumference. For a given structure, the following relationship can also be written:

$$R = \frac{\text{Re}\, v}{v} = \frac{v^2}{\text{Fr}\, g} = \sqrt{\frac{\text{Re}^2\, v^2}{\text{Fr}\, g}}. \qquad (6.29)$$

According to investigations by Fischerström (1955), stable, laminar flow can be expected in sedimentation basins when $R < 500$ and $\text{Fr} > 10^{-5}$. Substituting the above conditions into eq (6.29), the resulting R values and thus the corresponding basin dimensions will be such that they are in non-conformity with the basin dimensions generally used in practice. It is easily conceivable that the above conditions can only be met by dividing the basin volumes into segments by installing appropriate walls or plates. Namely, the Fr number can be increased either by increasing the flow velocity or by decreasing the hydraulic radius. However, increasing the velocities would be unfavourable as this increases the Re number and thus turbulence. Consequently, the only possible solution is to decrease the hydraulic radius, thereby increasing the Fr number and decreasing the Re number. For a given basin cross-sectional area A, the hydraulic radius can only be reduced by increasing the wetted circumference C_w. This unambiguously points to the above-discussed solution of splitting the basin volume into segments by installing separation walls.

The question of finding the most appropriate position of the space separation walls (the lamellas) remains open; their position might be horizontal, vertical or inclined.

Theoretically, horizontal splitting would be the most desirable as this yields the smallest settling/flotation depths (equalling the distance between the lamellas). From the practical point of view, however, an inclined arrangement of the lamellas is more justifiable, as this allows the relatively easy removal of substances deposited onto or floated up to the plates. The most unfavourable solution is vertical splitting of the space, since then the settling depth remains the total depth of the basin. Nevertheless, even in this case, the splitting up of the basin volume would yield stable laminar flow which is very favourable.

Fischerström (1955) carried out several investigations with lamellar settling basins — partly with new designs and partly with the conversion of existing facilities to lamellar structures. From these investigations, *Fig. 6.10* shows an approach that offers

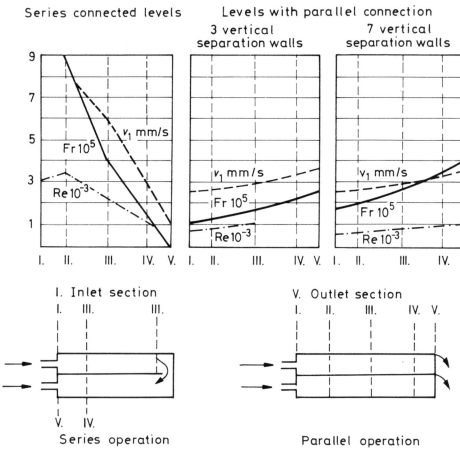

Fig. 6.10. Comparison of the hydraulic characteristics of lamellar settling basins, connected in series and parallel (after Fischerström 1955)

Table 6.1. Data of laboratory and prototype oil-trap experiments with single-lamella model of 4 m plate length and 10 cm distance between the plates

Discharge Q	Velocity v	Reynolds number Re	Froude number 10^{-5} Fr	Oil-trapping in % of the influent oil		Average oil droplet diameter			
				in labo- ratory	in pro- totype	in laboratory		in prototype	
						inlet	outlet	inlet	outlet
(l/min)	(m/min)	—	—	(%)		(micron)			
25	0.26	203	3.94	64	30	20.4	12.5	17.6	10.8
12	0.12	99	0.90	67	35	21.8	15.7	12.1	10.0
8	0.08	66	0.39	72	60	10.4	11.9	12.8	7.4

a comparison between series and parallel connected lamellar settling basins. The graphs show the variation of velocities and the respective Fr and Re numbers in cross-sections I—V. The parallel connection proved to be the more favourable one, in general. In the case of parallel connections of lamellar basins, all of the above-mentioned advantages will be achieved.

In *Table 6.1* and *Fig. 6.11*, laboratory and prototype measurement obtained by Szabó (1969) with lamellar oil traps are presented. This author proposes, for pro-totype applications, the use of a maximum plate length of 8—10 m and approx. 0.1 m distance between the plates. A guideline value of 0.12 m/min for upward flow velocity is given.

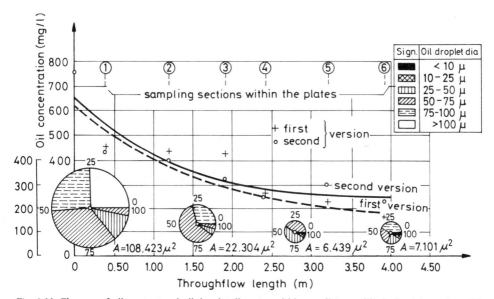

Fig. 6.11. Changes of oil content and oil droplet diameter within an oil trap with single plate pacing of 4 m and sewage flow rate of 25.1 l/min

11

6.9 Equalization ponds

Equalization ponds are frequently used as one of the steps in treating sewage with an oil and grease content (as a stage after flotation basins, tangential oil traps, etc.). These ponds allow further removal of floating contaminants due to the degrading of emulsions and to biological processes, etc. This might be an important aspect, since the effluents of traditional grease and oil traps can contain as much as 100 mg/l of lightweight phase. At the same time, however, the limit values of oil and grease content (organic solvent extract) of the effluents to be discharged into living waters and public sewers are 10 mg/l and 60 mg/l, respectively.

According to Mongajt and Rodziller (1958), an equalization storage time of 2-3 days is desirable for initial oil concentrations of 50—250 mg/l. The measurements by Lesenyei (1953) in Szőny (Hungary) also support this 3-day retention time. More stable emulsions, however, remained unchanged even after 45 days of storage.

6.10 Basic design data

For grease and oil traps, the design particle size is usually specified within the range of 0.25—0.3 mm. In some countries, the guidelines specify $d = 0.15$ cm. The average retention time is $t_c = 2$—3 min.

For oil traps, the maximum sewage flow velocity should be about 15 cm/s. The upward rising (buoyant) flow velocity w is approximately 1—5 mm/s. The density of the lightweight phase to be separated in the oil traps varies in the range of 0.8—0.9 g/cm^3.

According to German guidelines, a grease trap base-area of 0.25 m^2 is required for each 1.0 l/s of sewage discharge. To reach an efficiency higher than 90%, the following retention times are recommended (Sierp 1953).

Table 6.2. Characteristic design data for oil traps
(after Kropf 1957)

Oil type	Specific weight r_l (kp/dm^3)	Rising flow velocity w (m/h)	Specific area of the oil trap per unit flow rate m^2/litre/s
Petroleum ether	0.75	22.5	0.16
Petroleum	0.80	18.0	0.20
Light oil	0.85	13.5	0.27
Lubricating oil	0.90	9.0	0.40

Remark: Velocity w corresponds to $d = 0.25$ mm.

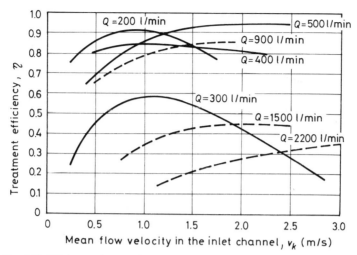

Fig. 6.12. Efficiency of tangential oil trap as a function of the inlet flow velocity

For sewage load $Q_v = 0.5$—9.0 l/s, min. 3 minutes;
For $Q_v = 10$—19 l/s, min. 4 minutes;
For $Q_v > 20$ l/s, min. 5 minutes.

Some characteristic design data for longitudinal-flow oil traps are summarized in *Table 6.2.* Swiss guidelines also specify $d_{min} = 0.25$ mm as the droplet diameter. In this case, an approx. 90% oil-separation efficiency is expectable. According to the size ranges of oil droplets, the following separation systems can be distinguished: a) rough, disperse or heterogenous system ($d > 500$ mμ); b) colloid systems ($1 < d(m\mu) < 500$); and c) solute systems ($d < 1$ mμ).

With respect to hydraulic investigations of tangential oil traps, the measurements by Lesenyei (1953) should be mentioned, and some of the respective results are shown in *Fig. 6.12.*

The curves specify relationship between the mean flow velocity v_k of the tangentially arriving raw water and the efficiency of oil removal. The third parameter is the discharge rate Q. The curves indicate that the optimum removal efficiency and the corresponding flow velocity v_k vary with the discharge rate.

In the case of gasoline traps a density of 0.85 g/cm^3 should be taken into account. A removal efficiency of at least 95% is required for such traps, and this can be achieved with an average residence time of about 3—6 minutes.

The allowable thickness of the floating substance (gasoline) is given by the Passavant Company as follows:
  for $Q_v < 2$ l/s, max 10 cm;

11*

for $Q_v = 2$—6 l/s, max 14 cm; and

for $Q_v > 6$ l/s, max 20 cm.

When the above thicknesses are exceeded during operation, the floating gasoline should be removed. In grease traps, the floating grease layer can expendiently be removed when its thickness reaches 1/5—1/10 of the water depth.

Air blowing can increase the efficiency of flotation facilities. For grease traps using blown air, Imhoff and Albrecht (1972) propose a specific rate of 0.2 m³ air/m³ sewage water, assuming a 3-minute retention time. The efficiency of flotation can be increased with an additional chlorine dosage to the blown air; a dosage corresponding to a 1.0—1.5 mg/l chlorine concentration in the influent is advisable, according to Randolf (1966).

7 Flotation basins

7.1 The principle of flotation

Flotation processes used in sewage-treatment technology are mainly used for removing hardly or not settleable solid or liquid particles or droplets from sewage water (mainly from industrial waste waters) and also for sludge thickening with flotation.

The efficiency of flotation can be increased by blowing air or gas into the sewage water, thus enhancing the ability of suspended particles to rise to the surface. Flotation can be efficient in removing grease, oil, fibrous substances (e.g., textile industry) and paper contaminants (paper and pulp industry). It is also a widely used procedure even in the case of substances having specific weights higher than that of water. In this context, it is mostly the waste waters of metal- and coal-processing industries that can be mentioned (ore enrichment, coal washing, etc.). More recently, flotation has also been used for sludge thickening in the field of activated sludge treatment and in that of chemical water treatment.

7.2 Conditions of linking and rising

When air bubbles are blown into the sewage water to be treated, the bubbles and the suspended particles might become linked to each other under certain physical conditions. Buoyant forces acting on the thus generated solid-air composite will be higher than the force of gravity acting on the solid particles. Consequently, the condition of rising is

$$\frac{\pi d^3}{6}(\gamma_f - \gamma_l) > \frac{\pi d_s^3}{6}(\gamma_s - \gamma_f), \qquad (7.1)$$

where γ_f, γ_l and γ_s are the specific weights of the liquid, the air and the solid, respectively. The other notations are shown in *Fig. 7.1*.

The density of an air bubble, containing vapour, can be calculated as

$$\varrho_l = 1.2929 \frac{273.1}{T} \left(\frac{p_b - 0.3783 p_g}{760} \right) \text{ (g/l)}, \qquad (7.2)$$

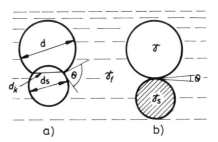

Fig. 7.1. Coupling of air bubble with solid or liquid particle

where T is the absolute temperature (°C), p_b is the barometric pressure (mm Hg) and p_g is the vapour pressure of moist air (mm Hg).

The force characterizing the coupling of air bubbles with solid particles is defined by surface tension and the angle of wetting Θ. As indicated by the Figure, the adhesion of air bubbles depends on the degree of wetting. For larger angle Θ the degree of wetting is smaller and the circumference πd_k of the coupling circle is larger. Consequently, the force $K = \pi d_k \sigma$ resulting from surface tension σ will also be larger.

The condition of the stable coupling of the two phases can thus be written as

$$\frac{\pi d_s^3}{6}(\gamma_s - \gamma_f) < \pi d_k \sigma , \tag{7.3}$$

which means that the weight of the solid particle in the liquid (i.e., the difference between gravity and buoyant forces) should be less than the force stemming from surface tension. It is seen from the Figure that particles with good wetting capability will not be carried to the surface by the rising air bubbles [version b)]. Therefore, the angle of wetting Θ, that defines diameter d_k, plays an important role in the efficiency of flotation. The angle of wetting can be improved by dosing with certain chemical agents, termed the "collectors" (some oil, certain salts, foam-generating agents, etc.).

7.3 Time of rising of the bubbles

For individual small bubbles (in the range of validity of Stokes' law) the time of buoyant rising can be derived, under hydrostatic conditions, in the following manner. Let us consider the starting conditions of a bubble as those of initial depth h_0, pressure p_0, diameter d_0 and volume V_0. After a certain time t, the values of the respective variables will be h, p, d and V. During buoyant rise the law of Boyl-Mariotte is considered valid, assuming an isothermic expansion: $p_0 V_0 = pV = \text{const.}$ Pressure p will be a composite of hydrostatic and atmospheric pressures:

$p = \gamma h + p_{atm}$. Therefore

$$V_0 p_0 = \frac{\pi d_0^3}{6}(h_0\gamma + p_{atm}) = Vp = \frac{d^3}{6}(h\gamma + p_{atm}),\qquad(7.4)$$

whence

$$\left(\frac{d}{d_0}\right)^2 = \left(\frac{h_0 + \dfrac{p_{atm}}{\gamma}}{h + \dfrac{p_{atm}}{\gamma}}\right)^{2/3}.$$

Considering a varying rising velocity $w = dh/dt$ and the validity of Stokes' law ($w = kd^2$), one obtains that

$$dt = \frac{dh}{w} = \frac{dh}{kd^2} = \frac{\left(h + \dfrac{p_{atm}}{\gamma}\right)^{2/3}}{k\left[d_0^2\left(h_0 + \dfrac{p_{atm}}{\gamma}\right)^{2/3}\right]}\,dh$$

and further

$$\int_0^t dt = \frac{1}{k\left[d_0^2\left(h_0 + \dfrac{p_{atm}}{\gamma}\right)^{2/3}\right]}\int_{h_0}^h \left(h + \frac{p_{atm}}{\gamma}\right)^{2/3} dh\,.$$

After integration, the time of buoyant rising of air bubbles is obtained as

$$t = -\frac{3}{5}\frac{h^{5/3} - h_0^{5/3}}{kd_0^2\left(h_0 + \dfrac{p_{atm}}{\gamma}\right)^{2/3}} = 10.8\,\frac{(h_0^{5/3} - h^{5/3})v}{gd_0^2\left(\dfrac{\gamma_s}{\gamma} - 1\right)\left(h_0 + \dfrac{p_{atm}}{\gamma}\right)^{2/3}},\qquad(7.5)$$

where

$$k = \frac{g}{18v}\frac{\gamma_s - \gamma}{\gamma}.$$

7.4 Flotation using dissolved and blown air

There are two ways of implementing flotation:

a) Air or gas is introduced into the water under pressure, and then, by relieving this pressure, very fine bubbles of uniform distribution are formed. The average diameter of the rising bubbles will be 80 micron and it ranges between 20—120 μ. The upward flow rate of the bubbles follows the Stokes' law. The actual rising velocity of suspended solids-air bubble composites will be 2.5—12.0 cm/min, and it increases with increase of the quantitative ratio of bubbles to solids. While rising, the bubbles become attached to the suspended solid particles of the water, thus decreasing the mean specific weight of the composite and inducing bouyancy.

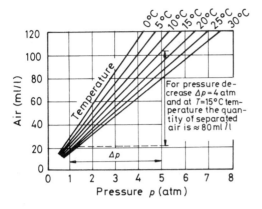

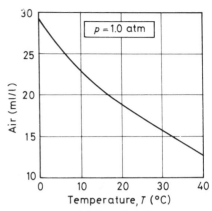

Fig. 7.2/a. Solubility of air in water as a function of the pressure and the water temperature (after Morse, in Fair et al. 1968)

Fig. 7.2/b. Solubility of air in water as a function of the water temperature, at atmospheric pressure (after Eckenfelder 1970)

b) By blowing air or gas into the water, bubbles can also be formed and the flotation process will be again similar to that of item a) above. The basic difference is that in this latter case the size of the bubbles is larger, even in the case of using the finest possible blower heads. The average bubble size is about 1000 micron (= 1.0 mm), if sufficiently high shear forces assure this.

The dissolution of air has been investigated by several researchers. It was proved that Henry's law is valid for dissolving oxygen and nitrogen in water. At the usually applied pressure of 1.5—3.5 atm, 35—40% of the total air quantity introduced dissolves. The water solubility of air is shown in *Figs 7.2/a* and *7.2/b*, as a function of pressure and temperature.

7.5 Design principles and relationships

Flotation systems can be further subdivided depending on whether the entire quantity of sewage is flotated or only part of it. The latter is the so-called recycling solution. The two solutions are illustrated in *Fig. 7.3*.

The most important design data for pressurized flotation facilities are: the pressure, the recirculation rate (if any), the suspended solids concentration of the sewage to be treated, the time of gas introduction and flotation (the retention time). The suspended solids content of the sewage decreases with increasing retention time, thus increasing the quantity of flotated substances. For the preliminary treatment of waste waters, a retention time of 20—30 min is usually sufficient. If thickening of sludge is the objective, then longer retention times should be allowed. In the pressure tank, a 1—3

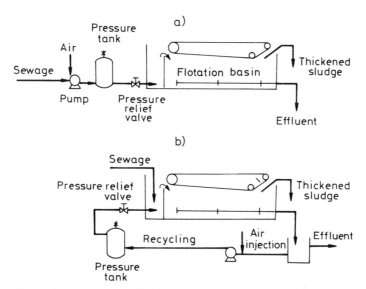

Fig. 7.3. Schematic arrangement of flotation systems; a) without recycling; b) with recycling
(after Eckenfelder 1970)

min retention time is sufficient, especially if mixing of appropriate intensity is also
applied. Flotation units can, in practice, be circular or rectangular. The collection and
removal of the floating scum (thickened sludge, oil, etc.) is carried out by various
mechanical collector and skimmer devices.

In the case of dissolved air or gas flotation operations, the efficiency of the process
can be calculated as

$$\frac{L_B}{L_s} = \frac{1.3 s_a R(P-1)}{Q_v C_1},\tag{7.6a}$$

where L_B/L_s is the air/solid ratio in the water to be treated; s_a is the air saturation
in cm^3/litre (at 1.0 atm); R is the recirculation rate (litre/day); P is the absolute
pressure (atm); Q_v is the water or sewage discharge (litre/day); and C_1 is the suspended solids concentration in the influent. If the effluent is not recirculated, then eq
(7.6a) can be simplified accordingly:

$$\frac{L_B}{L_s} = \frac{1.3 s_a (P-1)}{C_1}.\tag{7.6b}$$

The optimal value of the dimensionless ratio L_B/L_s was the subject of several
investigations, as this defines the efficient operation. *Figure 7.4* shows measurement
data (after Eckenfelder 1970) for different waste waters.

In the case of flotation systems using blown air or gas, some characteristic parameters can be calculated.

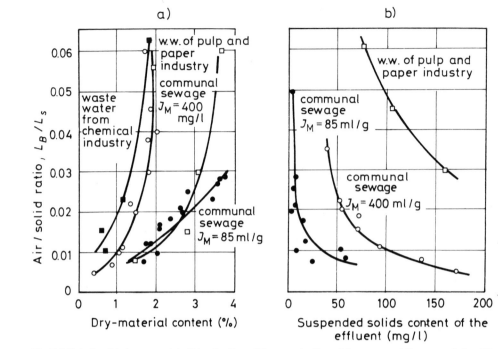

Fig. 7.4. Relationship between air/solid ratio, the solid concentration of the scum and the suspended-solids content of the effluent for different waste waters (after Eckenfelder 1970)

The number of bubbles present at a given point of time:

$$N_B = \frac{V_B}{\frac{\pi d_B^3}{6}}.$$
(7.7)

The number of bubbles generated during a unit period of time:

$$N_{B*} = \frac{Q_{air}}{\frac{\pi d_B^3}{6}}.$$
(7.8)

The boundary surface area of bubbles present in the water at a given point of time:

$$A = \frac{6 Q_{air}}{d_B} \frac{h_r}{w_B} \quad (m^2).$$
(7.9)

The boundary surface area of bubbles generated during a unit period of time:

$$A^* = \frac{6Q_{air}}{d_B} \quad (m^2/min).\tag{7.10}$$

Total bubble surface in a unit volume of water:

$$a = \frac{A}{V_R} = \frac{6Q_{air}}{d_B}\frac{h_r}{w_B}\frac{1}{V_R} \quad (m^2/m^3).\tag{7.11}$$

In the above relationships V_R is the volume of water in the flotation basin (m^3); V_B is the total volume of bubbles (m^3); d_b is the average bubble diameter (mm); w_B is the upward (rising) velocity of the bubbles (m/s); Q_{air} is the discharge rate of blown air (m^3/h); and h_r is the blowing depth of air during operation (m).

7.6 Basic design data

Some data of practical importance will be presented in this section. According to Fair, Geyer and Okun (1968) the bubble size in the case of blown air is approx. 1.0 mm, while that of the bubbles released from dissolved air is less than 0.1 mm. The required time of pressurizing to enrich the water with dissolved air is 0.5—3.0 min. According to Vrablik (1959), the bubble size of air released from the solution varies in the range of 0.03 mm—0.12 mm, while upward flow velocities can be well estimated on the basis of Stokes' law. The bubble-solid composite rises with a velocity of 2.5—12.5 cm/min. According to the proposals by Eckenfelder (1970), the desirable retention time of flotation basins is 20—30 minutes (when sludge thickening by flotation is desired, the upper limit should be extended). The surface-loading rate is 3.7—10.0 m/h.

The possibilities of treating various waste waters by flotation have been analysed by Stander and van Vuuren (1968). These authors found that flotation is, in many respects, more favourable than settling. First of all, flotation allows higher surface loading rates and can be used for treating denser and more concentrated sewages and sludges. It was found that the introduction of surface-active substances (such as alkilbensolsulphonate, ABS) can substantially improve the efficiency of the system. The authors used centrifugal pumps for dispersing air. They also developed a procedure for removing algae.

As was already mentioned, sludge thickening can also be performed by using flotation equipment. Some operational parameters in this respect will be presented on the basis of experiments by Kaeding (1962). First of all, it can be mentioned that the thickening of activated sludge by flotation can be done either with or without the addition of chemicals (surface-active agents). The latter solution is sometimes more favourable since the removal of chemical agents can be a source of further problems. Air is introduced into closed tanks at 3.0—3.5 atm. In the reactor space, air separates from water in the form of extremely fine bubbles in a uniform distribution. According to the results of pilot scale experiments, the separation of activated sludge from water can reach 90% efficiency during a retention time of about 15 minutes. The dry-material content of the floating sludge was 7%. The sludge can be removed from the surface by mechanical skimmers or — in the case of less coherent sludge or foam — by suction devices.

The specific air requirement per unit basin volume was 0.1 m^3/m^3h, while the energy demand was 0.15—0.25 kW/m^3.

For further information on practical applications, the works of Bratby and Marais (1974), Katz (1958) and also the publications of the Engineering Science Inc. can be referred to.

8 Filters

Filters of various types — which are equally both traditional and modern treatment facilities — are frequently used in the field of water technology. In sewage treatment their primary role is mainly in the field of tertiary or advanced treatment stages. Although filters do not belong exclusively to the category of mechanical waste water treatment facilities the dominating role of mechanical and physical effects during their operation suggest the desirability of discussing them in this Chapter.

8.1 Basic principles of filtration

First of all, a summary of the main terms used in sewage-filtration technology will be presented in order to provide a general framework for understanding the subsequent sections.

The *filtration process* consists of the discharge of a liquid with suspended solids content through a filter medium in order to remove the solid particles.

The *filtration cycle* = filtration period + washing period:

$$t_c = t_f + t_w . \tag{8.1}$$

The *porosity of the filter medium* is the ratio of the total pore volume (V_{pore}) to the total volume of the medium (including the pores):

$$n = \frac{V_{pore}}{V_{total}} = \frac{V_{total} - V_{medium}}{V_{total}} . \tag{8.2}$$

The characteristic porosity ranges are: gravel 0.3—0.4, sand 0.3—0.45, clayey soil 0.35—0.55, peaty soil 0.6—0.8.

The relative volume filled by the particles of the filter medium, that is, the particle volume per unit volume, is:

$$\frac{V_{medium}}{V_{total}} = 1 - n . \tag{8.3}$$

The *absolute porosity* can be calculated by eq (8.2), taking the total empty volume (V_{empty}) into account.

The *effective porosity* can be calculated by taking the interlinkable empty spaces and passages into account ($< V_{empty}$).

The *pore coefficient* is the ratio of pore volumes to the particle volume of the filter medium:

$$e = \frac{V_{pore}}{V_{medium}} = \frac{V_{pore}}{V_{total} - V_{pore}} = \frac{n}{1-n} .$$ (8.4)

The *effective particle diameter* (according to Hazen 1892) is the 10% value of the particle size distribution curve

$$d_e = d_{10} .$$ (8.5)

Further characteristic particle diameters are:
a) the 50% value of the particle size distribution curve, d_{50};
b) the maximum particle diameter, d_{max};
c) the diameter corresponding to the inflexion point of the distribution curve, d_m.
The coefficient of uniformity:

$$U = \frac{d_{60}}{d_{10}} .$$ (8.6)

The specific surface area is the total surface area of particles in a unit volume of the medium:

$$S = \frac{6(1-n)}{d_e} \quad (m^2/m^3) .$$ (8.7)

In practice, it is desirable to consider the shape coefficient ψ, by dividing the above expression by ψ (the value of ψ is 0.85 for sand and 0.70 for anthracite).

The number of particles in a unit volume of the filter is

$$N_p = \frac{1-n}{\dfrac{d_e^3 \pi}{6}} = \frac{S}{d_e^2 \pi} \quad (pc/m^3) .$$ (8.8)

The shape coefficient characterizes the deviation of the shape of the particles of the medium from that of a sphere:

a)
$$\psi = \frac{\text{surface area of a sphere having the same volume as that of the particle}}{\text{surface area of the particle}}$$ (8.9)

b)
$$\frac{\text{surface area of the particle}}{\text{volume of the particle}} = \frac{\alpha_d}{d} = \frac{\alpha_D}{D_p} .$$ (8.10)

The value of shape coefficient α_d depends on the definition of particle diameter d. If it is defined as diameter D_p of the smallest sphere that can contain the particle, then the value of α_p of particles of various geometric forms will be as follows (Kovács 1972a):

sphere: $\alpha_p = 6$; cube: $\alpha_p = 10.4$; octaeder: $\alpha_p = 10.4$; tetraeder: $\alpha_p = 18$; quartz sand: $\alpha_p = 7$—11.

The *pore diameter* is the diameter d_p of a cylinder having the same specific surface area as that of the pore.

The volume between the particles is

$$V_{empty} = n V_{total} = \frac{n}{1-n} V_{medium} \quad (m^3).$$ (8.11)

The *hydraulic radius is defined* as the ratio of the empty spaces between the particles to the surface area of the particles:

$$R_h = \frac{V_{empty}}{f_p} = \frac{n V_{total}}{f_p} = \frac{n}{1-n} \frac{V_{medium}}{f_p} \quad (m).$$ (8.12a)

This can also be written as

$$R_h = \frac{n}{1-n} \frac{d}{6},$$ (8.12b)

since for a single particle $V_{medium} = 4r^3\pi/3$; $f_p = 4r^2\pi$; $V_{medium}/f_p = r/3 = d/6$.

The *filtration (seepage) velocity* or rate is defined as the ratio of the rate of flow Q_v loaded onto the filter to the filter cross-sectional area A_f perpendicular to the direction of the flow:

$$v_f = \frac{Q_v}{A_f} \quad (m/h).$$ (8.13)

To calculate local mean flow velocities, the actual cross-section of flow should be taken as

$$v_t = \frac{v_f}{n} \quad (m/h).$$ (8.14)

The *hydraulic gradient* is defined as the head loss corresponding to the unit thickness (l) of the filter medium:

$$I = \frac{dp_x/\gamma}{d_x}; \quad I = \frac{\Delta p_x/\gamma}{H_f} = \frac{h}{l},$$ (8.15)

where

$h = \frac{\Delta p_x}{\gamma}$ and H_f is the thickness of the filter layer.

8.2 Hydraulic principles

Water flowing within the filter medium can be considered as seepage flow, and the respective hydraulic principles should be used in the hydraulical design of the filters. Due to the irregular arrangement of pores and passages within the filter medium, the

seepage pathway of water is of an extremely complex character. As a first approximation, the expressions derived for laminar pipe flow can be used for the hydraulic
analysis. By increasing the filtration rate, the transient and turbulent flow regimes can
also be obtained. The critical value of the Reynolds number characterizes the transition from laminar to turbulent flow.

In filtration technology the Blake number is more frequently used; this is actually
a modified Reynolds number:

$$\text{Re} = \frac{vl}{v} = \frac{v_t R_h}{v} = \frac{\dfrac{v_f}{n} \dfrac{dn}{6(1-n)}}{v} = \text{Bl} \tag{8.16a}$$

or

$$\text{Bl} = \frac{dv_f}{6(1-n)v} = \frac{v_f}{Sv}, \tag{8.16b}$$

where d is the particle diameter (mm); v_f is the filtration rate (m/h); n is the pore
volume and v is the kinematic viscosity (mm^2/s).

For rapid filters, the value of the Blake number varies in the range of 0.2—0.55.
When calculating the Blake number, the characteristic velocity of the Reynolds
number is considered as $v_t = v_f/n$ and the hydraulic radius is considered as having
following form $R_h = dn/6(1-n)$.

The basic law of laminar flow in porous media was defined by Darcy in 1856, by
relating the seepage velocity to the hydraulic gradient as

$$v_f = k_D \frac{h}{l} = k_D \frac{\Delta p}{\gamma l} = k_D I. \tag{8.17}$$

Darcy's law holds for low filtration rates only, when flow velocities are in the
laminar domain. The hydraulic permeability coefficient K_D includes the effects of
all factors that affect the relationship $v_f = f(I)$. In Hazen's relationship, K_D is expressed in terms of the effective diameter d_e as

$$v_f = 115.7 \, d_e^2 \, I, \tag{8.18}$$

where $d_e = d_{10}$ in cm and v_f is obtained in cm/s. The above relationship refers to a
water temperature of 10 °C and to particle sizes of 0.1—3.0 mm, while the uniformity
condition should be $U < 5$.

Flow in a filter medium can be approximated by relationships derived for pipe flow.
Thus, the following well-known formula can be written:

$$h = \frac{\Delta p}{\gamma} = \zeta \frac{l}{d} \frac{v_f^2}{2g}, \tag{8.19}$$

where h is the head loss in cm; l is the thickness of the filter medium, cm; d is the
particle diameter (cm); v_f is the filtration rate in cm/s and ζ is the resistance coefficient. The value of resistance coefficient ζ can be calculated by the following expressions:

a) in the laminar range (Re < 10):

$$\zeta = 2000 \left(\frac{v_f d}{v} \right)^{-1} ; \qquad (8.20a)$$

b) in the turbulent range (Re > 300):

$$\zeta = 94 \left(\frac{v_f d}{v} \right)^{-0.16} ; \qquad (8.20b)$$

c) the intersection of the above two curves defines the critical Reynolds number in the transitional range (10 ≤ Re ≤ 300):

$$Re_{crit} = 38 .$$

Rearranging eq (8.19) and assuming laminar flow, seepage velocity v_f is obtained for a temperature of 10 °C, as

$$v_f = 74.7 \, d^2 I . \qquad (8.21)$$

The deviation from Hazen's relationship eq (8.18) is due to differences in defining the particle size.

In the case of laminar flow, the expressions presented in the literature for calculating the hydraulic gradient can, generally, be traced back to the Koženy–Carman (in Kovács 1972a) formula. For example, Kroupa (1960) proposes the following expression:

$$I = \frac{h}{l} = 193 \, \frac{v(1-n)v_f}{gn^3 d^2} , \qquad (8.22)$$

which is essentially the Koženy–Carman equation (cm/s and cm units should be substituted).

According to this author diameter d should be calculated as

$$d = \sqrt{\frac{6 G_p}{\pi \gamma N}} , \qquad (8.23)$$

where G_p is the total mass (g) of particles in a sample and N is their number in the same sample.

In an in-depth hydraulic analysis it is more appropriate to use dimensionless numbers. According to Rouse (1958) the following general dimensionless relationship of gravity, friction and inertia forces should be sought:

$$\frac{h}{l} = f(Re, Fr, n) . \qquad (8.24)$$

In the laminar range the effect of inertia can be neglected (provided that the assumption of uniform linear flow is valid). In this case, the Fr/Re ratio is used in the general formula (eq 8.24).

8.3 Transport processes and mechanisms

In order to establish contact between the suspended-solid particles and the particles of the filter media certain transport processes should take place, in the course of which the suspended particles will cross the flow pathways (the streamlines). If they do not, then the suspended particles would move along the flow pathways and pass the particles of the filter medium. Consequently, the experimental fact that particles much smaller than the pore size can also be removed by filtration could not be justified. *Figure 8.1* clearly illustrates the orders of magnitude of suspended particles in the vicinity of a pore.

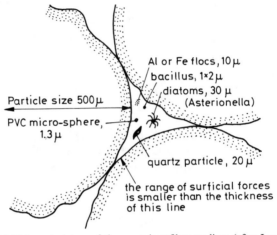

Fig. 8.1. Enlarged picture of the pores in a filter medium (after Ives 1969)

In discussing the mechanisms of filtration, first the transport processes should be underlined. The relevant literature discusses several transport mechanisms that occur in filter media. Some of the most important of these will be briefly discussed below, while some of them are illustrated by *Fig. 8.2*.

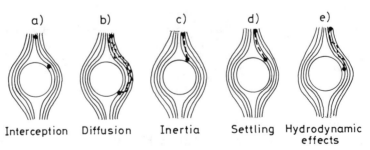

Fig. 8.2. Basic transport mechanisms (after Ives 1969)

Straining effect

If the size of the suspended solids in the water to be filtered is relatively larger than — or of the same order of magnitude as — the pore size of the filter medium, then the upper thin layer of the filter bed will retain the suspended particles, thus to a large extent defining the effectiveness of the filter. The upper 3—5 cm layer of the filter will soon become clogged, thus not allowing the lower layers to take an effective part in the filtration process. The same situation might occur when smaller suspended particles are building up, due to coagulation and flocculation processes, an arch over the pore openings. In the case of such suspensions, it is then more desirable to apply surface strainers or, with appropriate conditioning (for example clarification) to separate the suspended solid fraction which causes rapid clogging before the filtration process.

Interception

According to *Fig. 8.2a*, a suspended particle might be directly intercepted by a filter particle if it travels with a current that passes right around that particle of the filter medium. This effect can be characterized by the following relationship:

$$i = \frac{e_s}{d}, \tag{8.25}$$

where e_s is the representative size of the suspended particle. This mechanism differs from the straining effect in that interception can retain much smaller particles than straining. The particles brought mechanically into contact with the surface of a sand filter medium will be retained there due to adhesion.

Diffusion

Due to the Brownian motion of the molecules, the pathway of suspended particles less than a micron in size might be altered randomly. This is illustrated by *Fig. 8.2b*. In the case of particles less than 1.0 millimicron in size, friction forces might reduce the diffusion effect, thus decreasing the so-called free length of travel to one or two times the particle diameter. The diffusion mechanism can be characterized by the dimensionless Peclet number ($v = v_f$):

$$Pe = \frac{vd}{D}, \tag{8.26}$$

where D is the Stokes–Einstein diffusion constant;

$$D = \frac{kT}{3\pi\eta e_s}, \tag{8.27}$$

in which k is the Boltzmann constant and T is the temperature in $°K$.

12*

Inertia effect

According to *Fig. 8.2c*, the pathway of motion of suspended particles transported by the flow might deviate from the current due to inertia forces and they thereby come into contact with the surface of the filter medium. On the basis of the Navier–Stokes equation, the following dimensionless number can be derived for this situation:

$$\frac{\varrho_l e_s^2 v}{18 \eta d} = \frac{1}{18} \cdot \frac{e_f}{d} \cdot \frac{\varrho_l e_f v}{\eta} = \frac{1}{18} i \,\mathrm{Re}. \tag{8.28}$$

Settling

If the density of suspended solids differs significantly from that of the flowing medium and the particles are relatively large, then the settling process, as illustrated by *Fig. 8.2d*, might occur. In this case, the pore voids can be considered as small settling basins in which particulate matter leaves the flow lines and settles out due to gravity. The dimensionless number characterizing this process is

$$\frac{(\varrho_l - \varrho) e_f^2 g}{18 v} = \frac{1}{18} \frac{\mathrm{Re}}{\mathrm{Fr}}, \tag{8.29}$$

that leads to Stokes' settling equation. There is a certain relation between eq (8.29) and the inertia parameter equation (8.28), although the difference is essential, namely, in eq (8.29) the role of inertia is eliminated from the Re/Fr ratio and only gravity and friction forces play a role. Consequently, the ratio Re/Fr decreases with the increasing velocity, as opposed to the value of i Re that increases with increasing v.

Hydrodynamic effect

In discussing filtration mechanisms, the hydrodynamic effect is the one that stems from rotation (moment of momentum) of the suspended particles. Due to a non-uniform velocity gradient distribution, to an irregular particle shape and also to the deforming character of the particles, the rotation of particles will result in forces such as will cause deviation of the pathway of particles from the current, thus increasing the probability of their becoming attached to the surface of the filter medium (see *Fig. 8.2/e*). The characteristic Reynolds number of this process can be written as

$$\mathrm{Re} = \frac{\varrho \, dv}{\eta}. \tag{8.30}$$

The role and value of this number depends, obviously, on the definition and value of the physical variables of the right-hand side of eq (8.30). A special case of this Re number is the Blake number as defined by eq (8.16).

The effect of orthokinetic flocculation

Orthokinetic flocculation can also be considered as a filtration mechanism, since flocculation due to the velocity gradient — which plays a basic role — might increase the size of the particles, thus increasing the efficiency of removal. This mechanism is characterized by the dimensionless Camp number:

$$Ca = Gt_s \tag{8.31}$$

that can be calculated by considering the following relationships. The velocity gradient G is obtained as

$$G = \left(\frac{\varrho\, g v I}{n\eta}\right)^{1/2} \tag{8.32}$$

and

$$t_s = \frac{h_f n}{v_f}. \tag{8.33}$$

Substituting the Koženy–Carman formula for hydraulic gradient I, eq (8.31) becomes

$$Ca = Gt_s = 13.4\,\frac{(1-n)}{n}\,\frac{h_f}{d}. \tag{8.34}$$

According to eq (8.34), the Camp number and thus the extent of flocculation depends on the filtration rate and on the temperature. Ives (1969) considers consideration of the term $Ca - xC_0$ more justifiable (C_0 is the initial concentration of suspended solids content C and h_f is the distance within the filter layer).

The joint effect of transport mechanisms

The above-discussed effects will, eventually, act simultaneously. Depending on the fluid, its suspended solids content and on the properties of the filter medium and also on the conditions of operation, some of the effects might be negligible, while other become dominating. As general information, it may be stated that the role of diffusion will be unambiguously significant at particle diameters smaller than one millimicron. For particles larger than 10 microns, settling will play the larger role (if their density is greater than that of the water). The role of interception also increases with the increase of particle size.

The general mathematical form for the combined effect of all mechanisms can be written as

$$A = f\left(i;\ Pe;\ i\,Re;\ \frac{Re}{Fr};\ Re;\ Ca\right), \tag{8.35}$$

where the values of the various dimensionless numbers depend on the interpretation of the various physical quantities (e.g., the Reynolds number can also be calculated with different physical units). The actual form of eq (8.35) can be derived on the basis of experimental measurements for a given case. For example, Ison and Ives (1969) obtained the following relationship for the removal of 2.5—10.0 micron caolinite (clay) particles:

$$A = \text{const}\ i^{-2/3}\left(\frac{Re}{18\,Fr}\right)^{1.3} Re^{-2.7}, \tag{8.36}$$

whence the initial value of filter coefficient λ_f is

$$\lambda_0 = \text{const}_1 \, \frac{\eta^{1.4} e_f^{0.3}}{d^{1.4} v^4} \,. \tag{8.37}$$

For micro-size PVC spheres, Sholji (in Ives 1969) derived the following empirical relationships:

$$A = \text{const} \, \frac{d}{P_e} \tag{8.38}$$

and

$$\lambda_0 = \text{const}_2 \, \frac{1}{\eta^2 v} \,. \tag{8.39}$$

The above relationships can be used for actual calculations and they also show the effects of various variables. It should be emphasized, however, that the validity of the above relationships is restricted to the conditions under which the respective experiments were made. The number of variables affecting the filtration process is so great and the effects of the different mechanisms are so complex that actual design relationships can be derived only on the basis of experimental measurements.

8.4 The mechanisms of attachment and detachment

In addition to the above-discussed transport mechanisms the efficiency of the filtration process depends basically on the mechanisms of attachment and detachment of the suspended particles with and from the surfaces of the filter medium. Due to the effects of transport mechanisms, the suspended particles come into contact with these surfaces to which they might become attached (fixed) due to the effects of certain forces. In certain situations, the already attached particles might become loosened from the surfaces detachment. For example, by increasing the filtration rate, an equilibrium condition might be achieved at which the same number of particles will be attached and detached during a unit period of time.

The process of attachment is essentially defined by physicochemical and molecular forces. First of all, the electrical dual-layer effect should be emphasized. As is known, the dual electrical layer that develops on the surface of particles suspended in water results in the generation of repulsion and attraction forces, depending on the value of the electrokinetic zeta potential. *Figures 8.3* and *8.4* are shown to demonstrate the essence of this zeta potential. Measurement results have indicated that in water most of the solid and suspended particles possess negative zeta potentials. This means that the dual electrical layer tends to prevent the process of attachment. The extent of this effect is defined, among other things, by the presence of dissolved solids (salts). This feature provides the theoretical basis of chemical dosage for treatment purposes.

The second, also basically important, effect is due to the London–Van der Waals force, i.e., the attraction between atoms and molecules. In reality, however, the

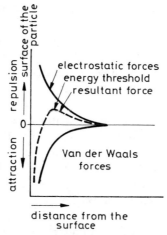

Fig. 8.3. Distribution of surficial forces

different forces act simultaneously. The joint effect of the dual electrical layer and the London–Van der Waals forces was investigated by Ives and Gregory (1967). On the other hand, the joint effect of hydrodynamic forces and the London–Van der Waals force was analysed by Heertjes, Lerk and Mackrle (in Ives 1969).

Some authors found that hydrogen binding of water molecules might also play an important role. Nevertheless, the effects that can be achieved by dosage of chemical

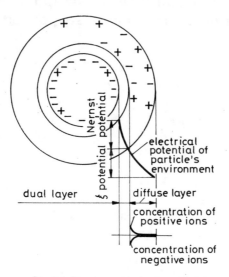

Fig. 8.4. Development of zeta potential

agents will play a more important role. Various polymers provide "bridges" between suspended solid particles and the sand particles of the filter medium, thus increasing filtration efficiency. Polyelectrolites — and even traditional clarification agents, such as aluminium sulphate and iron chloride — decrease the repulsion effects of the dual electric layer, by shifting the value of the zeta potential towards the negative direction. It is to be noted that non-ionic and anion-active polymers might also enhance the process of attachment.

Transport processes and the phenomenon of attachment are, essentially, reversible processes. This explains the process that takes place during filter washing, which can be considered a detachment process. It is obvious that the process of detachment is of lesser importance during the filtration period, although inappropriate design or shock-like loading might decrease the treatment efficiency. The process of detachment plays a more important role during filter-washing operations. The role of the detachment process has been investigated by Mints et al. (1966). Other authors, however, did not consider this process as significant within the filtration process. It can, however, be stated that the dosage of chemical agents will also basically affect the process of detachment. Flocs and aggregates will, as a function of their strength, determine the process of detachment.

In our opinion, the processes of attachment and detachment should be analysed in their interrelationship, as will also be demonstrated by discussing some of the relevant mathematical models below.

8.5 Some important mathematical models for describing the temporal and spatial distribution of suspended solids

The description of spatial and temporal processes taking place within filter media can be made essentially by two basic relationships;

a) the equation of continuity, and
b) a dynamic equation for the description of velocities.

The equation of continuity

In our case, the equation of continuity expresses the assumption that the quantity of contaminants removed from the suspended phase equals the mass retained by the filter medium. This can be expressed in a simplified form as

$$-v_f \frac{\partial C}{-\partial h_f} = \frac{\partial \sigma_f}{\partial t}.$$

(8.40)

Taking the time variation of concentration C also into consideration, it can be written that

$$-v_f \frac{\partial C}{\partial h_f} = \frac{\partial \sigma_f}{\partial t} + (n - \sigma_f) \frac{\partial C}{\partial t}, \qquad (8.41)$$

where σ_f is the volume of particles retained in a unit volume of the filter medium.

The velocity equation

The velocity equation describes the rate of suspended-solid removal, as well as the variation of the concentration with depth. The most widely used relationship is the differential equation derived by Iwasaki (1937) for slow filters:

$$-\frac{\partial C}{\partial h_f} = \lambda_f C, \qquad (8.42)$$

which states that the quantity of suspended solids retained by a unit depth of the filter layer is proportional to the local concentration C. Solving the above differential equation for initial conditions $h_f = h_0$; $C = C_0$ and $\lambda_f = \lambda_0$ it is obtained that

$$C/C_0 = \exp(-\lambda_0 h_f). \qquad (8.43)$$

The initial condition is that λ_0 refers to the filtration coefficient of a clean filter bed.

Mints et al. also considered the process of detachment:

$$-\frac{\partial C}{\partial h_f} = \lambda_0 C - \frac{\alpha_f \sigma_f}{v_f}, \qquad (8.44)$$

where the first and second terms on the right-hand side refer to the mechanisms of attachment and detachment, respectively. α_f is the parameter of detachment. For $\alpha_f = 0$, eq (8.44) simplifies into eq (8.42).

Variables affecting filter coefficient λ_f

Filter coefficient λ_f implies the joint effect of an extremely large number of parameters. Thus, the value of λ_f varies with the filter design and with the conditions of operation. There are many formulae proposed by different authors for the calculation of the value of λ_f under different conditions. One of the most general relationships was elaborated by Ives (1969):

$$\frac{\lambda_t}{\lambda_0} = \left(1 - \frac{\sigma_f}{\sigma_u}\right)^x \left(1 + \beta \frac{\sigma_f}{\eta}\right)^y \left(1 - \frac{\sigma_f}{n}\right)^z, \qquad (8.45)$$

where σ_u is the ultimate value of σ_f, when the operation of the filter becomes ineffective, i.e., it no longer retains suspended solids. It follows from eq (8.45) that for $\sigma_f \to \sigma_u$, $\lambda_f \to 0$, a fairly obvious result.

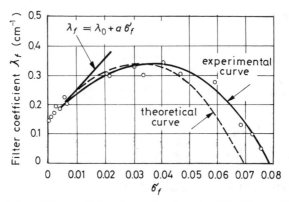

Fig. 8.5. Variation of filter coefficient λ_f as a function of σ_f (after Fox an Cleasby 1966)

Equation (8.45) is in conformity with some of the other equations proposed by different authors for the special cases of filtration:

the Iwasaki–Stein formula:

$$\lambda_f/\lambda_0 = 1 + \beta \frac{\sigma_f}{n} \tag{8.46a}$$

and

$$\lambda_f = \lambda_0 + \alpha\sigma_f, \tag{8.46b}$$

when the parameters of general equation (8.45) are $x=0$; $y=1$; $z=0$ (x, y and z are coordinates). An equation from Ives (1969):

$$\lambda_f = \lambda_0 + \alpha\sigma_f - \frac{b\sigma_f^2}{n - \sigma_f}, \tag{8.47}$$

for which the parameters are $x=y=z=1$. Equation (8.47) allows the joint consideration of the processes of attachment and detachment. The curve described by this equation can be compared to measurement results, as illustrated by the experimental evaluation of *Fig. 8.5.*

Mackerle's relationship

$$\frac{\lambda_f}{\lambda_0} = \left(1 + \beta \frac{\sigma_f}{n}\right)^y \left(1 - \frac{\sigma_f}{n}\right)^z, \tag{8.48}$$

when $x=0$;

The Shektman, Heertsjes–Lerk relationship is

$$\frac{\lambda_f}{\lambda_0} = 1 - \frac{\sigma_f}{\sigma_u} \tag{8.49}$$

if $x=y=1$ and $z=0$;

A formula from Maroudas:

$$\frac{\lambda_f}{\lambda_0} = 1 - \frac{\sigma_f}{n} \tag{8.50}$$

if $x=1$ and $y=z=1$.

Substituting proportionality $\alpha_f \propto C$ from eq (8.44) conditions $x=1$ and $y=z=0$ are again obtained, as was demonstrated by Ives (1969).

In respect to the above-discussed mathematical models it should be noted that although they were both theoretically and experimentally well founded, their practical application might be a source of inaccuracy. The reason is that it is fairly difficult to render numerical values to σ_f and σ_u. The engineering design concept is better suited to relationships that define the changes of the filtration coefficient in terms of the operating variables such as:

$$\lambda_0 = \frac{\text{const}}{v_f^\alpha d^\beta \eta^\gamma}, \tag{8.51}$$

where the values of the empirical constants vary with the technological conditions within the following ranges:

$$\alpha = 0.7 \div 4; \quad \beta = 1 \div 3; \quad \gamma = 2 \quad \text{or} \quad \gamma = -1.4.$$

If the size e_f of the particles of suspended solids is also taken into account in eq (8.51), then

$$\lambda_0 \propto e_f^\delta, \tag{8.52}$$

where the value of δ varies (according to the experimental results of Yao 1970) between -0.5 and 1.6.

It is to be noted that some examples of using eq (8.51) have already been presented in discussing the joint effects of transport mechanisms (sign $\propto$ refers to proportionality).

The mathematical relationships describing the distribution of concentration with

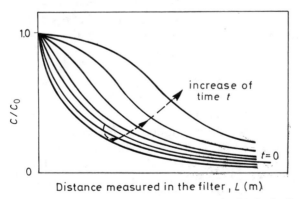

Fig. 8.6. Time and space variation concentration of solids in the filter

I. Mechanical treatment

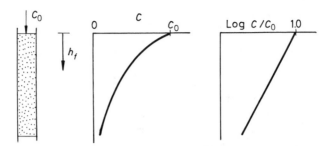

Fig. 8.7. Logarithmic distribution of concentration in a filter of uniform particle dimensions
(after Ives 1969)

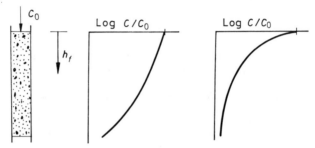

Fig. 8.8. Distribution of concentration in a filter of non-uniform particle dimensions (after Ives 1969)

depth can also be verified experimentally. Taking samples from different depths at
different times the series of curves shown in *Fig. 8.6* can be drawn. The Figure is
presented after Ives (1969), who pointed out the time variation of the shape of these
curves. *Figure 8.7* shows an exponential concentration distribution, in accordance
with eq (8.43). *Figure 8.8* suggests that the exponential distribution does not remain
valid in all cases. The curve, plotted on a semi-logarithmic scale, cannot be approximated by linear lines. Consequently, the generalization of eq (8.4) becomes justifiable.
A possible approach is eq (8.44) derived by Mints et al. (1966).

8.6 Description of head conditions

In addition to concentration, distribution the other important way of characterizing the operation of filters is the description of the head conditions. Before giving the
details, it is necessary to discuss some of the basic hydraulic relationships.

It is known that in seepage flow Darcy's law prevails in the laminar domain that
defines the linear relationship between seepage velocity and the hydraulic gradient

$$v_f = k_D I. \tag{8.53}$$

In filtration technology, the Koženy–Carman relationship is generally used, which is essentially Darcy's law with a more detailed expression of the coefficient of permeability:

$$I = \frac{\partial h}{\partial h_f} = K \frac{\eta S^2 v_f}{\varrho g n^3} = 36 K \frac{\eta (1-n)^2 v_f}{\varrho g n^3 d_e^2 \psi^2},$$ (8.54)

where K is the Koženy constant, and the following substitutions can also be applied:

$$\varrho = \frac{\gamma}{g} \quad \text{and} \quad v = \frac{\eta}{\varrho}.$$

Combining eqs (8.53) and (8.54), it is obtained that

$$k_D = \frac{1}{K} \frac{\gamma n^3}{\eta S^2} = \frac{1}{36 K} \frac{\gamma n^3 d_e^2 \psi^2}{\eta (1-n)^2}.$$ (8.55)

Porosity n and specific surface area S will both change during the filtration process, due to the clogging effect of the retained suspended particles. Consequently, the permeability properties will also change, and this should be duly considered. In the relationship proposed by Ives (1969) the changes in the ratio I/I_0 can be expressed as

$$\frac{I}{I_0} = 1 + (2\beta + 1) \frac{\sigma_f}{n} + (\beta + 1)^2 \left(\frac{\sigma_f}{n}\right)^2 + (\beta + 1)^3 \left(\frac{\sigma_f}{n}\right)^3 + \dots .$$ (8.56)

Similarly, Ives (1969) proposes, on the basis of the Koženy–Carman formula, the following expression:

$$\frac{I}{I_0} = \left(\frac{1 - n_0 + \sigma_f}{1 - n_0}\right)^{4/3} \left(\frac{n_0}{n_0 - \sigma_f}\right)^3.$$ (8.57)

If $\sigma_f \ll n$ then, according to eq (8.56) the hydraulic gradient is linearly proportional to the retained specific mass quantity σ_f:

$$I = I_0 + \text{const } \sigma_f.$$ (8.58)

The value of the total head loss changes both in time and space. For the time variation, some authors propose the following linear relationship:

$$h = h_0 + \text{const } t,$$ (8.59)

where h_0 is the initial total head loss across the filter. According to our experience, this linear relationship can be valid for relatively narrow ranges and only for certain filter technologies.

In addition to the time variation of head conditions, valuable design information might be obtained from knowledge of the pressure distribution within the filter column. The pressure diagram of Michau (1951), as shown in *Fig. 8.9*, is the usual way of illustrating the vertical hydraulic head distribution in a filter. Line *MO* marks the hydrostatic pressure line. At the initial point $t_0 = 0$ of starting up the filter, the

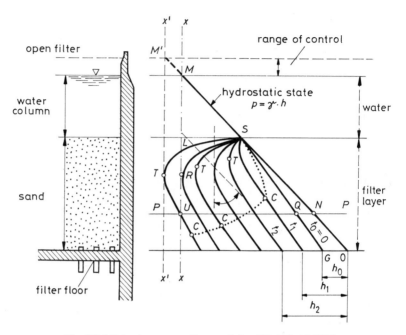

Fig. 8.9. Michau's pressure diagram (after Wiederhold 1954)

hydrodynamic pressure line *SG* will represent the head conditions. At this point in time, the filter is still clean and free of clogging. With the progression of time, the head distribution curve will be shifted to those marked by $t_1, t_2 \ldots$ on the Figure. There might be a point in time when the pressure distribution curve touches the *x—x* axis (in point *R*) and this means that at the filter depth defined by this point, the hydraulic head equals the atmospheric pressure. Continuing the filtration, the head-distribution curve will intersect the *x—x* axis at two points (points *U* and *L*, where $\pm p/\gamma = 0$). There will be depression (vacuum) in the filter layer falling between these two points, a condition that might result in the formation of bubbles from the dissolved gases present in the water (this might also induce clogging). To eliminate this situation, the water depth should be increased. The required additional water depth is obtained by shifting axis *x—x* to the *x'—x'* position (which is the vertical line tangential to the head-distribution curve) and finding point *M'* as the intersection of the hydrostatic line *MO* with axis *x'—x'*. *M'* marks the required new water level.

From Michau's diagram, both the head loss and the actual head values can be obtained at any filter layer depth and at any point in time. For example: for filter level *P—P*, *UQ* is the pressure at time point $t_0 = 0$, while *QN* marks the head loss due to the transition from the hydrostatic to the hydrodynamic state. At the bottom level of the filter, the total head loss is found as h_0, h_1, h_2 for times $t_0, t_1, t_2 \ldots$, respectively, while the actual pressure is represented by the remaining distance to axis *x—x*.

The Figure also allows the determination of the depth to which contaminants penetrated and were retained in the filter. The extent of clogging is represented by the angle between the tangential line of a selected point on the pressure distribution curve and the line *SG* (marked with dashed lines in the Figure). Above the dotted line connecting the points marked by *C*, the head-distribution lines are curved, while below this line they are linear and parallel with line *SG*. In the case of correctly designed filters, point *C* of the head-distribution curve should, after the prescribed filtering time, approach the lower level of the filter bed. If this point is much above this lower level, then the filter was not sufficiently utilized, since the curve represented by points *C* marks the penetration depth of the contaminants. The head-distribution curves also indicate that the head loss is the highest in the upper layers of the filter medium, since clogging is the most intensive in this range. At a selected point in time, the lowest head can be measured at point *T*, which is the nearest point of the curve to axis *x—x*. According to our experience, the above linear relationship will be valid for certain filtration technologies only.

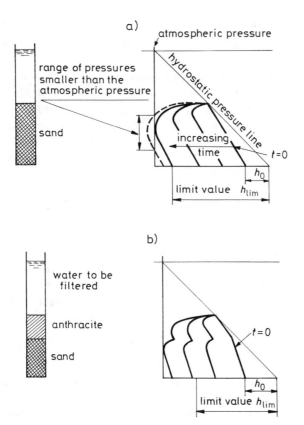

Fig. 8.9/a–b. Michau's diagram in single (a) and dual-layer (b) filters

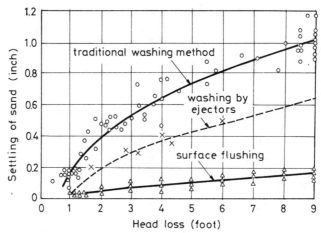

Fig. 8.10. Relationship between head loss and the settling of sand (after Baylis, Gullans and Hudson, in Ives 1969)

In *Fig. 8.9/a–b* comparison is made between the pressure diagrams of single- and double-layered filters.

Finally, a relationship between head losses and the settling (shrinking) of a sand filter bed is shown in *Fig. 8.10.* Head conditions will, eventually, affect the compression of the different layers of the filter bed. Using the conventional washing methods, only the upper clogged layers of the filter cannot be fully eliminated in all cases and the pressure on this surface results in compaction and shrinking of the filter bed during the filtering period. This effect can be reduced by using various methods (cleaning by ejectors, surface flushing, etc.).

8.7 Washing of filters

During the operation of the filter, a point in time is reached when either the head loss or the suspended solids concentration of the effluent will exceed a limit value. At this point, regeneration of the filter becomes necessary. The periods of filtration and washing — within a single filter operation cycle can be determined experimentally. Washing is carried out by a current of water of appropriate intensity with a direction opposite to that of the filtration. For most filter types, the wash water flows upward, carrying away the impurities from the pores of the filter medium. During washing the original pore volume n will be increased to n_e, depending on the extent of expansion. The relationship between washing rate, pore volume and head loss is illustrated by *Fig. 8.11.*

The degree of expansion can be interpreted as follows. The force resulting from the friction head loss of the upward flowing water can be balanced by the total weight of the particles (taking buoyant forces also into account):

$$\varrho g A_f h_e = (\varrho_l - \varrho) g A_f (1 - n_e) H_e, \tag{8.60}$$

where h_e is the head loss corresponding to the given expansion, and H_e is the total depth of the expanded filter layer.

Hydraulic gradient I can be expressed from the above balance equation as

$$I = \frac{h_e}{H_e} = (1 - n_e) \frac{\varrho_l - \varrho}{\varrho}, \tag{8.61}$$

whence the total head loss h_e, needed for washing, can be calculated. If the filter medium is composed of particles of different densities, then the head loss can be calculated as

$$h_e = \sum_{i=1}^{n} (1 - n_{en}) \frac{\varrho_{ln} - \varrho}{\varrho} p_n H_e, \tag{8.62}$$

where ϱ_{ln} is the density of particles in the n-th filter layer, n_{en} is the porosity of the n-th expanded layer and p_n is the fraction of the depth of the n-th layer.

The measure of expansion can be derived (as a kind of efficiency) as follows:

$$E = \frac{H_e - H_f}{H_f} = \frac{H_e}{H_f} - 1 = \frac{1 - n}{1 - n_e} - 1 = \frac{n_e - n}{1 - n_e}. \tag{8.63}$$

The value of E can also be expressed as a percentage (by multiplying by 100). If the intensity of washing is also considered, then the ratio H_e/H_f can be defined as

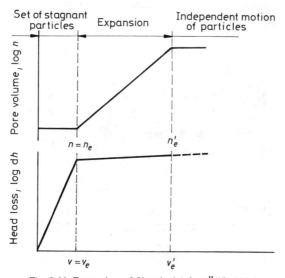

Fig. 8.11. Expansion of filter bed (after Öllős 1970)

13

follows:

$$\frac{H_e}{H_f} = \frac{1-n}{1-\left(\dfrac{q}{v_s}\right)^a}, \tag{8.64}$$

where $q = v_w$ is the rate of washing, and v_s is the settling rate of the particles; a is an empirical constant that varies between 0.2 and 0.3, depending on the hydraulic conditions and on the shape of the particles. Equation (8.64) can also be generalized for multi-layered filters as

$$\frac{H_e}{H_f} = (1-n) \sum_{i=1}^{n} \frac{p_n}{1-\left(\dfrac{q}{v_w}\right)^a}. \tag{8.65}$$

The mechanism of detachment is of basic importance with respect to the flushing operation. The opinions of various authors as to whether the detachment of impurities from the sand-particle surfaces is due to collision of the particles or to friction forces of the flowing media are rather diverse. We believe that both forces play an important role, which depends on the degree and character of clogging. In this context the rate of filtration also plays an important role. Collisions of particles can be effective when the filter bed is in fluidized condition. In the case of larger expansion, the particles will be further from each other and the probability of collision reduced.

Various methods are used to increase the efficiency of washing:

a) washing combined with air blowing;
b) washing by water jets;
c) washing with the application of mechanical devices (surface scrapers) in order to losen the filter surface;

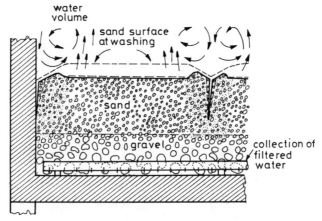

Fig. 8.12. Effect of clogged areas during filter washing (after Baylis, Gullans and Hudson, in Ives 1969)

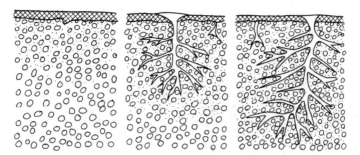

Fig. 8.13. Scheme of channelling in a filter bed (after Baylis, Gullans and Hudson, in Ives 1969)

d) more recently, and in the case of continuously operating filters, regeneration is carried out by hydraulically mixing up the filter bed.

The above methods also allow of a reduction of the quantity of wash-water. In the case of efficient washing, the quantity of wash-water may not exceed 3% of the filtrate produced.

The most frequently applied washing rates are in the range of 15—50 m/h. In Hungary, they are usually in the lower part of this range. Air-blowing rates are usually in the range of 50—70 m/h. In order to prevent the carrying away of filter material, devices serving for the conveyance of wash-water (collection troughs or flumes) should be installed at an appropriate distance from the surface of the filter bed, taking expansion also into consideration. The distance usually applied in Hungary is about 70 cm.

Finally, some illustrative examples of the conditions prevailing in a filter bed during washing will be presented. *Figure 8.12* indicates that in the vicinity of heavily clogged locations, washing may be ineffective and expansion is reduced. *Figure 8.13* shows a completely clogged upper filter layer with the development of "channel" pathways. It can be seen that clogging and channelling are gradual processes. In this case, washing should remove the upper clogged layer as well. Finally, *Figs 8.14* and

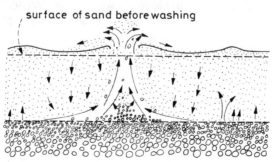

surface of sand before washing

Fig. 8.14. Sand currents at the beginning of washing (after Baylis, Gullans and Hudson, in Ives 1969)

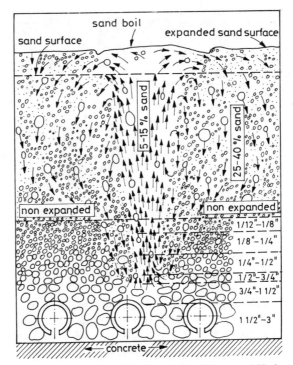

Fig. 8.15. Movement of sand in a filter bed (after Baylis, Gullans and Hudson, in Ives 1969)

8.15 demonstrate the movement of sand during the washing process. The boil-like sand flows, within the filter bed, are well indicated by these figures.

8.8 Hydraulic investigations of ultra-rapid filters

The author has recently carried out theoretical studies and experimental measurements on the hydraulic and technological aspects of increasing filtration rates. Some important details of the results (published elsewhere) will be presented below.

On the basis of theoretical considerations and experimental results, the following relationships between filtration rate v_f and hydraulic gradient I have been elaborated:

$$v_f = v_{max} \frac{I}{I + I_f} \quad \text{and} \quad I = I_f \frac{v_f}{v_{max} - v_f}. \tag{8.66a—b}$$

The above relationships represent a hyperbolic relationship between v_f and I, with the model parameters being v_{max} and I_f. These two parameters involve, in a lumped manner, the effects of all geometric and physical variables that affect the filtration process (particle size, pore volume, viscosity, density, temperature, etc.). It is justifiable to find a technological meaning for parameter I_f. First of all, it can be stated that I_f is a dimensionless quantity of the hydraulic gradient type. If in eq (8.66a) the hydraulic gradient equals I_f, then

$$v_f = v_{max}/2, \quad \text{or} \quad \frac{v_f}{v_{max}} = 1/2. \tag{8.67a}$$

In other words, this means that I_f is a hydraulic gradient at which the filtration (seepage) rate is half of v_{max}. Considering two other possible extreme conditions:

$$v_f = \frac{v_{max}}{I_f}, \quad \text{or} \quad \frac{v_f}{v_{max}} = \frac{I}{I_f} \quad \text{for} \quad I \ll I_f \tag{8.67b}$$

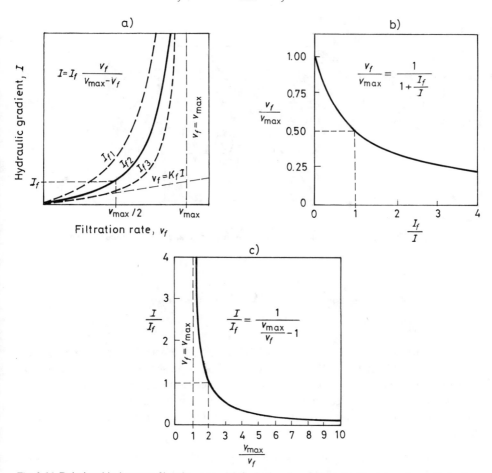

Fig. 8.16. Relationship between filtration rate and the hydraulic gradient, in dimensioned and dimensionless forms

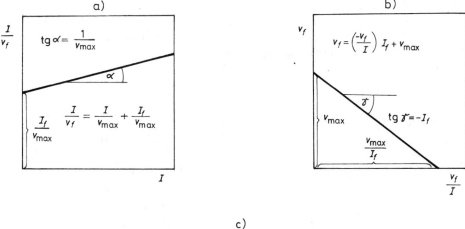

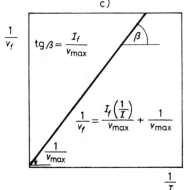

Fig. 8.17. Linearized relationship between filtration rate and the hydraulic gradient

and

$$v_f = v_{max}, \quad \text{or} \quad \frac{v_f}{v_{max}} = 1.0 \quad \text{for} \quad I \gg I_f. \tag{8.67c}$$

The conditions represented by eqs (8.67a—c) are demonstrated, in a dimensionless form, in *Fig. 8.16*. In *Fig. 8.17*, a graphical method is presented, in three alternatives for calculating the values of v_{max} and I_f, when linearizing eqs (8.66a—b).

The question arises as to what relationship between the above equations and the known hydraulic laws can be found. In this context, three different cases might be discussed:

a) in the case of $I = I_f$, the respective equation [eq (8.67a)] shows an analogy with the Hagen–Poisseuille formula, stating that in a laminar pipe flow the mean flow velocity equals half of the maximum flow velocity measured in the pipe axis;

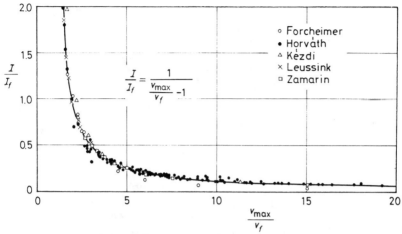

Fig. 8.18. Dimensionless relationship between filtration rate and the hydraulic gradient

b) If $I \ll I_f$, eq (8.67b) represents Darcy's law, and the value of permeability coefficient k_D is obtained as

$$k_D = \frac{v_{max}}{I_f} = k_f; \qquad (8.67d)$$

c) Then if $I \gg I_f$, eq (8.67c) is the case, showing an analogy with a velocity profile of turbulent state.

Expanding eq (8.66b) into a series with the McLaurin method, a further relation can be determined:

$$I = \frac{I_f}{v_{max}} v_f + \frac{I_f}{v_{max}^2} v_f^2 = a v_f + b v_f^2. \qquad (8.68)$$

Equation (8.68) shows analogy with the parabolic formula of Forcheimer (1930). An extended version of the latter is obtained by expanding the series with the higher-power terms. At low velocities, when terms of powers higher than I can be neglected, eq (8.68) will again yield Darcy's law.

In order to verify the above-discussed relationships, experimental data of several authors, along with our own, were analysed in a combined way. *Figure 8.18* shows the relationship between filtration rate and the hydraulic gradient in terms of the above dimensionless units. It can be unambiguously stated that the theoretical curve fits well to the measurement data of different authors, thus providing a verification of the model. Similarly, in *Figs 8.19/a* and *8.19/b* the theoretical approach presented above is shown together with Forcheimer's and Zamarin's formulae (in Kovács 1972a), respectively.

Summarizing, the following conclusions and proposals can be drawn and made:

a) Relationships were derived between filtration rate and the hydraulic gradient for the conditions of the high-rate sand filters frequently applied in the field of water

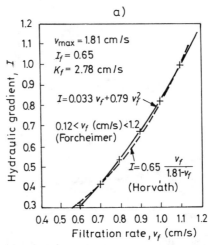

Fig. 8.19/a. An approach to the Forschmeier relationship 1. Hydraulic gradient, 2. Filtration rate, v_f

technology. The development of these relationships was justified by the discrepancies found between measurement data and the values obtainable by the Koženy–Carman formula in the case of filtration rates exceeding 20—25 m/h.

b) For $I \ll I_f$ the relationship is converted into Darcy's formula. Another extreme case is when $I \to \infty$ for which the relationships yield $v_f \to v_{max}$ that differs from the hydraulic theory. Nevertheless, the high ranges of I are of no practical importance in filtration technology.

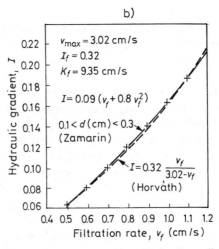

Fig. 8.19/b. An approach to the Zamarin relationship 1. Hydraulic gradient, 2. Filtration rate, v_f

c) The validity of the derived relationships were experimentally proved, using literature data along with our own measurement data. It can be stated that the theoretical curves fit well with the measurement data within the accuracy required for technological applications.

d) A new calculation method is proposed for the case $Re_f > 4$, while for $Re_f < 4$ the Koženy–Carman relationship yields appropriately accurate results.

e) The dimensionless form of the derived relationships allows a certain generalization of the achieved results.

f) Further investigations and measurements are needed for a more detailed hydraulic interpretation of the parameters v_{max}, I_f and k_f. The relationship between these parameters and the geometric (particle size, porosity, etc.), hydraulic (laminar, turbulent and transient conditions) and technological (water quality, chemical dosage, etc.) parameters of the filtration process should be quantified.

8.9 Hydraulic design of the structural elements of filters

The hydraulic design of channels and pipelines associated with filter equipment can be made on the basis of the conventional hydraulic principles. Nevertheless, some structural elements of the filters call for certain social considerations from the hydraulic design point of view. Such elements are the filter bed, the drain system used for collecting the filtrate, the throughs collecting the wash-water, etc. Some of these elements will be discussed below in the light of practical examples.

Design of the filter bed

In the case of filtrate drainage with a grid of perforated pipes, headloss h_v of the drainage system can be calculated, according to Kulskiy (1964), as

$$h_v = 0.25 + 3 \frac{v_{col}^2}{2g} - 9 \frac{v_{col,r}^2}{2g} + 18 \frac{v_{col,h}^2}{2g}, \qquad (8.69)$$

where v_{col} is the mean velocity at the beginning of the collector drain pipe (m/s); $v_{col,r}$ is the mean velocity at the beginning of the shortest forking pipe drain, and $v_{col,h}$ is the mean flow velocity at the beginning of the longest forking drain pipe; while g is the acceleration of gravity (9.81 m/s²).

The guideline value for the distance between filter heads can be calculated with the following formula:

$$l = \frac{100}{\sqrt{n}} \quad (cm), \qquad (8.70)$$

where n is the number of filter heads in a unit area (m²) (approx. 30—100 pc/m²).

The number of drain-pipe junctions is obtained for a collector drain length of L as

$$x = \frac{2L}{l}.$$ (8.71)

The total slot surface area of slotted filter heads is

$$f = \frac{Aw}{700\sqrt{2gh}} \quad (m^2),$$ (8.72)

where A is the filter area (m^2), w is the washing rate ($l/s\,m^2$), and h is the hydraulic resistance of the openings of the collector drain system (m). Thus, the number of slots to be applied is

$$k = \frac{f}{nA}.$$ (8.73)

The total surface area of the holes of perforated filter floors is

$$f = \frac{Aw}{10^3 \mu \sqrt{2gh}} \quad (m^2),$$ (8.74)

where h is the required pressure head in the distribution pipe line (m) and μ is the discharge coefficient of perforated openings ($\mu \approx 0.62$). The values of mean velocities are: 1.0—2.0 m/s for branching pipes and 0.8—1.5 m/s for distribution pipes.

Calculations related to filter washing

The quantity of wash-water needed for washing filter columns can be calculated by the following formula:

$$P_m = \frac{NW}{qt} 100 \quad (\%),$$ (8.75)

where N is the number of filter units; W is the water quantity required for a single washing of the filter; q is the planned discharge rate of water consumption (m^3/h); and t is the time between two filter washings (hours).

Carrying out the collection and conveyance of wash-water by troughs, their cross-sectional area A_v should be calculated as

$$A_v = \frac{Aw}{1000nv} \quad (m^2),$$ (8.76)

where n is the number of troughs and v is the mean flow velocity in the thoughs (m/s).

Surface washing, which is frequently applied in some countries, is carried out either by a fixed pipe system or by revolving sprinklers. For revolving sprinklers, the outlet jet velocity is 20—30 m/s and the intensity of washing is 0.50—0.75 $s \cdot m^2$. For fixed pipe systems, the intensity shall be 3.0—4.0 $s \cdot m^2$. The required head is 3—5 atm, to be provided by appropriate pumps.

8.10 Summary of the most important expressions used in the hydraulic-technological design of filters

Filtration rate:

$$v_f = \frac{Q_v}{A_f} = \frac{H_f}{t_f} \quad (\text{m/h}). \tag{8.77}$$

Actual flow velocity within the pores:

$$v_a = \frac{v_f}{n} = \frac{Q_v}{A_f n} = \frac{H_f}{t_f n} \quad (\text{m/h}). \tag{8.78}$$

Volume of water filtered during a filtration cycle t_f:

$$V_{tv} = Q_v t_f = A_f v_f t_f \quad (\text{m}^3). \tag{8.79}$$

Solids retained by the filter during a unit period of time:

$$L_f = Q_v (C_0 - C_e) = A_f v_f (C_0 - C_e) \quad (\text{g/h}). \tag{8.80}$$

Weight of retained solids per unit filter surface:

$$l_f = \frac{L_f}{A_f} = v_f (C_0 - C_e) \quad (\text{g/m}^2 \cdot \text{h}). \tag{8.81}$$

Energy consumption of filter operation:

$$E = \frac{\Sigma Q_v p \Delta t}{\eta_{\text{pump}}} \quad (\text{kWh}). \tag{8.82}$$

Treatment, filtration efficiency:

$$f = 100 \frac{C_0 - C_e}{C_0} \quad (\%). \tag{8.83}$$

Notations used in the above equations:
A_f filter surface area; H_f depth of filter (m); Q_v sewage water load on the filter (m^3/h); t_f calculated average detention time in the filter (h); p pressure (kg/cm^2); C_0, C_e suspended solids concentration of the influent and effluent, respectively (g/litre); n pore volume; η_{pump} pump efficiency.

8.11 Selected hydraulic questions concerning land (soil) filtration

A special field of filtration is the utilization of natural soil as the filter medium. According to experience, suspended solids are retained by soil at high rates. For example, Krone, Orlob and Hodgkinson (1958) analyzed the changes in bacteria counts during the filtration process, both theoretically and experimentally. Their manner of thinking can be summarized as follows.

The kinetics of filtration is determined by a first-order reaction kinetics as proposed by Iwasaki (1937) [see also eq (8.42)]:

$$-\frac{dn}{dL} = Fn,\qquad(8.84)$$

where n is the number of bacteria in a unit volume of the filter medium; L is the filtration length and F is the filter coefficient.

For initial conditions $n = n_0$ at $L = 0$ the solution of eq (8.84) is obtained as

$$\frac{n}{n_0} = e^{-FL}.\qquad(8.85)$$

Denoting the number of bacteria retained by a soil by N_t, the theorem of the conservation of matter yields that:

$$\frac{dN_t}{dL} = n\frac{dV}{dL} + V\frac{dn}{dL},\qquad(8.86)$$

as the changes of N_t equal the difference between the bacteria counts measured at the beginning and at the end of a filter of unit length. V is the fluid volume filtered. If the soil particles are uniform and the flow is steady, then $dV/dL = 0$, and

$$\frac{dN_t}{dL} = V\frac{dn}{dL} = -FVn = FVn_0 e^{-FL}.\qquad(8.87)$$

Of the experimental results of the above authors, the following should be emphasized. Very good removal efficiency of bacteria (95—98%) was achieved — with two-phase seepage — when using water infected by *E. Coli* and with presettled sewage. The filter medium was sterilized soil of 0.063—1.1 mm particle size. With three-phase seepage, the removal efficiency was low.

Table 8.1. Allowable surface loading rate of soil filters
(after Imhoff and Albrecht 1972)

Effective particle diameter d_e (mm)	Max. surface loading rate T_f (m/h)
0.2	0.8— 2.1
0.3	2.1— 4.2
0.4	4.2— 8.4
0.5	8.4—12.5

Remark: The above data refer to periodical loading
with the assumption of an unclogged filter
surface.

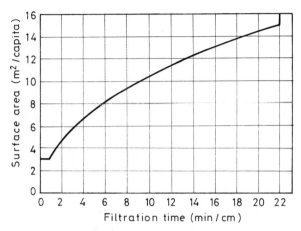

Fig. 8.20. Graph for calculating the required filtration surface area (after Pallasch and Triebel 1967)

From the practical point of view, the determination of saturation depth L_s may also be of importance. Investigations by Hall (in Kovács 1972a) gave information in this context:

$$L_s = \frac{V - V_s}{V_s} \frac{1 - (F_1' + F_2')}{F_1' + F_2'},$$ (8.88)

where

$$F_1' = 35a \left(\frac{D}{d}\right)^{3/2}$$

and

$$F_2' = 0.1 \frac{wp}{kI},$$ (8.89a—b)

where the new notations are: V_s is the volume of water filtered through a unit prism of soil; D is the diameter of solids retained by the filter; d is the particle size of the filter medium; a is a factor of proportionality (for sand filters $a = 0.038$—0.063); p is the useful pore volume; w is the settling velocity (from Stokes' law) $kI = v_f$ is the laminar seepage velocity (as in Darcy's law).

The design of soil filters should primarily be based on practical experience. Guideline data for this purpose are presented in *Table 8.1. Figure 8.20* is a guide for designing smaller plants. Here the filter surface means the total area of the seepage ditch. The required specific ditch length can be calculated thereby (as a guideline value, a 10—20 m ditch length per capita can be used).

II. Chemical treatment

9 Coagulation and flocculation equipment

Coagulation and flocculation are, essentially, transport processes associated with the linkage, floc formation of colloidal substances and particles. The thus generated larger sized particles can then be removed by settling, clarification or filtration. The rate of linkage — the kinetics of flocculation — is, consequently, a basic parameter in designing the respective equipment. Flow conditions play a significant role as well. The grouping of the various sub-processes of coagulation can also be made on the basis of hydraulic considerations. The time variation of coagulation is affected mainly by two factors: a) the Brownian motion and b) the interaction of particles. This latter is determined, partially, by the hydraulic conditions.

Before flocculation, chemical reactions (hydrolisis and precipitation) take place depending upon the effects of chemical dosage. Such reactions take place instantaneously and are followed by an equally rapid coagulation sub-process, the perikinetic (or molecular) coagulation, which is dominated by the Brownian motion. In this process the upper size limit of the particles involved is approx. 10 micron. Perikinetic coagulation is a colloid-chemical type process in which the surficial charge, the Zeta potential and the colloid stability, etc., play a deterministic role. The aggregation of particles is the result of diffusion transport processes.

Orthokinetic (or flow-induced) coagulation is the second stage of the coagulation process, in which physical processes such as the settling rate of particles, flow velocities and mixing conditions (laminar and turbulent flow conditions) play the dominating role. In the process of orthokinetic coagulation — in accordance with the principles of modern water-treatment technology — the notion of velocity gradient plays the primary role. (The terms "perikinetic" and "orthokinetic" were first used by Wiegner in 1911.) This chapter deals with the basic relationships of coagulation and flocculation and summarizes the basic hydraulic aspects of designing the respective treatment facilities.

9.1 Zeta potential and the electrophoretic motion (*EM*)

Coagulation and flocculation processes are associated with the chemical destabilization of colloids. Destabilization can be achieved by reducing the surficial charge (i.e., the Zeta potential) thus eliminating the repulsive forces between particles and inducing conditions in which suspended particles tend to coagulate and flocculate — and to settle out subsequently. The Zeta potential is a characteristic measure of the surficial charge of particles, the explanation thereof is illustrated by *Figs 8.4* and *9.1*. Numerically, the Zeta potential is defined by the well-known Helmholtz–Smoluchowski relationship that was derived on the basis of hydraulic and electrostatic considerations.

The dominating force (apart from electrical ones) acting on a colloidal particle that moves within an electrical field of forces due to the effects of potential gradient, is the force of friction. The transport equation, in the direction x of the motion, can be written (after Sennett and Olivier 1965) as

$$E\omega \, dx = \eta \frac{d^2 v}{dx^2} \, dx. \tag{9.1}$$

The Poisson equation of electrostatics should also be considered in the form of:

$$\nabla^2 \psi = -\frac{4\pi\omega}{\varepsilon}, \tag{9.2}$$

where v is the velocity of ions and particles moving upon the effect of an electrical potential — the electrophoretic moveability ($cm \cdot s^{-1}$); E is the potential gradient (Volt/m, $cm^{-1/2} \cdot g^{1/2} \cdot s^{-1}$); ω is the charge per unit volume ($cm^{-3/4} \cdot g^{1/2} \cdot s^{-1}$); ψ is the electrical double-layer potential at distance x from the surface (Volt $\cdot cm^{1/2} \cdot g^{1/2} \cdot s^{-1}$); ε is the dielectrical constant of the medium; η is the dynamic viscosity of the medium (poise).

From eqs (9.1) and (9.2), two characteristic dimensionless numbers can be derived by similarity transformations (Horváth 1970)

$$\pi_1 = \frac{E\omega l^2}{\eta v} \quad \text{and} \quad \pi_2 = \frac{\psi \varepsilon}{\omega l^2}. \tag{9.3a—b}$$

Assuming that the product of π_1 and π_2 is constant in systems of similar type:

$$v = \text{const} \, \frac{\varepsilon E \psi}{\eta}.$$

The special value of ψ is Zeta potential ζ. Thus, using the $\psi = \zeta$ substitution the Helmholtz–Smoluchowski relationship is obtained (for $\text{const} = 1/4\pi$):

$$v = \frac{1}{4\pi} \frac{\varepsilon \zeta E}{\eta} = k_E E; \quad \zeta = 4\pi \frac{\eta v}{\varepsilon E}. \tag{9.4a—b}$$

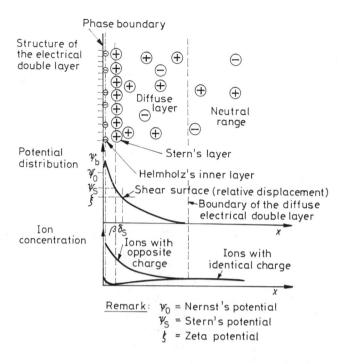

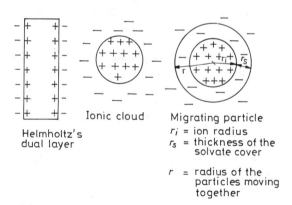

Fig. 9.1. Interpretation of the electrical double layer and the Zeta potential (after Popp et al. 1973)

The basic relationship of eq (10.4) has actually been obtained by solving differential eqs (9.1) and (9.2) on the basis of the theory of similarity. Regarding coefficient k_E as one analogous to the permeability coefficient of seepage processes, eq (9.4a) can be considered as an analogy to Darcy's law where the hydraulic gradient I is replaced by potential gradient E. The same relationship can be obtained for fluid motion in a capillary tube, induced by an electrical current.

K_E is the electroozmotic permeability coefficient. According to Eckenfelder (1970) the following simplified relationships can be used for practical purposes:

$$\zeta = \frac{113,000}{\varepsilon} EM; \qquad \zeta = 12.85\,EM \qquad\qquad (9.5a\text{---}b)$$

at temperateure $T = 20\ °C$, where $EM = v$ (micron/s · Volt/cm) and the value of ζ is obtained in mV units. It is to be noted that for the colloidal suspended solids of surface waters, the value of ε ranges between -8 mV and -20 mV.

9.2 Velocity gradient and energy dissipation

According to the original ideas of Camp and Stein (1943), the relationship between the velocity gradient and the energy dissipation plays a determining role in coagulation and flocculation processes. Further research results indicate that the concept of designing treatment facilities on the basis of the velocity gradient might be used for the characterization of hydraulic processes in a uniform way.

The velocity gradient that develops at various points in a flowing medium is the result of the turbulence of the liquid when introduced into the structure plus the effects of mixing that is applied within the structure (in addition to the shear stresses of laminar flow). The relationship between dissipated energy and the velocity gradient can be described — according to the above authors — by considering the shear stresses that correspond to laminar flow.

Considering an elemental fluid volume of side lengths dx, dy and dz, velocity gradient dv/dz will be the result of shear stress $\tau = \eta\, dv/dz$, i.e., of shear force $(\eta\, dv/dz)dx\, dy$. The work done by the shear force during a unit period of time is

$$\eta \left(\frac{dv}{dz}\right)^2 dx\, dy\, dz$$

and finally the work per unit volume and unit time, i.e., the dissipated energy of the laminar flow is obtained as

$$d = \eta \left(\frac{dv}{dz}\right)^2 = \eta G^2, \qquad\qquad (9.6)$$

whence

$$G = \sqrt{\frac{d}{\eta}} = \sqrt{\frac{D}{\eta V}}, \qquad\qquad (9.7)$$

where G is the velocity gradient that can be approximated, for a finite volume, from its time-averaged value $G = \sqrt{G^2}\ (s^{-1})$; $d = D/V$ is the energy dissipated in a unit volume of fluid during a unit period of time $(\mathrm{m} \cdot \mathrm{kg/h} \cdot \mathrm{m}^3)$; D is the dissipated energy, by flow, during a unit period of time (m kp/h); V is the volume of the fluid (m^3) and η is the dynamic viscosity of the flowing medium.

Rearranging eq (9.7), and introducing the calculated average residence time $t_s = = V/Q$:

$$\frac{Q}{V} = \frac{1}{t_s} = \frac{\sqrt{D/\eta V}}{Gt_s} \tag{9.8a}$$

or, in dimensionless form:

$$Gt_s = \frac{V}{Q} \sqrt{\frac{D}{\eta V}} = \frac{\sqrt{DV/\eta}}{Q} = \frac{Q_d}{Q}. \tag{9.8b}$$

Considering, as an example, a basin of length L and useful volume of $V = LBH$, then the work done by the force of friction — the energy dissipated — during a unit period of time can also be expressed by the following formula, in terms of hydraulic radius R and mean flow velocity v:

$$D = Q \Delta p = Q\varrho h_v = Q\varrho \lambda \frac{L}{4R} \frac{v^2}{2g} = \frac{\lambda LBH}{8R} \varrho v^3, \tag{9.9}$$

further, by substituting $V = LBH$ and expressing the energy dissipated in a unit volume of water, it is obtained that

$$d = \frac{\lambda}{8R} \varrho v^3. \tag{9.10}$$

Combining eqs (9.7) and (9.10).

$$G = \sqrt{\frac{v^3 \lambda}{8 v R}}. \tag{9.11}$$

It is to be noted that for the calculation of G, the values of D or d can be calculated in practice by the energy introduced, or by taking the pressure head and the geodetical elevation differences into account. The desirable solution will be determined by the character of the structure.

The above considerations indicate that the energy dissipation, as defined by the relationship $d = D/V = G^2$ and the velocity gradient G (and thus the Camp number), play a determining role in characterizing the flow conditions. In eq (9.8b) the Camp number Ca can be interpreted as the ratio of two characteristic rates of flow — that of the rate of flow Q_d, induced by the dissipated energy, to influent flow rate Q. Further, in eq (9.8a) hydraulic load (and thus loadability) Q/V depends, in the case of Ca = const, not only on V and t_s, but also on D and η. The optimal values of G and Gt_s can be obtained experimentally for various technological operations. For example, for flocculent substances and assuming a mean residence time $t_s = 10 \div 30$ minutes, the favourable ranges are (according to Fair, Geyer and Okun

14*

1968) $30 < G\,(\mathrm{s}^{-1}) < 60$; $10^4 < \mathrm{Ca} < 10^5$. The optimal velocity gradient can be interpreted also for settling operations, since above certain G values shear stresses might reduce floc size and thus the settling rate.

According to the research results of Camp and Stein (1943) the optimal value of G, in laboratory flocculation experiments is $20\ \mathrm{s}^{-1}$, while the disaggregation of flocs starts at $G = 40\ \mathrm{s}^{-1}$. This result suggests that the dimensions of the experimental equipment also play a role.

On the basis of summarizing American experiences, Camp and Stein proposed $G = 20$—$74\ \mathrm{s}^{-1}$ as design values for flocculation basins. The respective favourable ranges of the Camp number Ca would be $\mathrm{Ca} = 2.3 \times 10^4$—$21.0 \times 10^4$.

The above line of thought can be extended also to include turbulent flow conditions. The respective handbooks (for example the well-known book Fair, Geyer and Okun 1968) consider eq (9.7) to be valid for turbulent conditions as well, with the remark that it means dynamic viscosity in laminar flow only, while in turbulent flow it is a quantity of identical units but of a different meaning. A relationship similar to eq (9.7) has been derived by Levics (1958) on the basis of the diffusion theory. Essentially, it may be stated that the Camp–Stein formula can be considered valid for any flow conditions of Newtonian fluids.

In turbulent flow the relationship describing shear stress τ will be expanded to include additional turbulent shear terms (as opposed to the Newton law):

$$\tau = \eta\frac{d\bar v_x}{dz} + \varrho l^2\left(\frac{d\bar v_x}{dz}\right)^2 = \eta\frac{d\bar v_x}{dz}\left(1 + \frac{\varrho l^2}{\eta}\frac{d\bar v_x}{dz}\right),\qquad(9.12)$$

where $\bar v_x$ is the mean velocity in direction x and l is the mixing length of Prandtl.

In turbulent flow, stresses due to viscosity might be neglected in many situations, thus obtaining

$$\tau = \varrho l^2\left(\frac{d\bar v_x}{dz}\right)^2 = \eta^*\frac{d\bar v_x}{dz},\qquad(9.13)$$

where η^* is the virtual dynamic viscosity.

In eq (9.13) η^* — and thus l — are functions of the velocity gradient. It should be stressed once again that the substitution $\eta^* = \eta$ is allowable in the case of laminar flow only.

According to the investigations of Ives (1969) in sludge-curtain type clarifiers, eq (9.7) can be used in the case of turbulent flow as well. Equations (9.6) and (9.7) can be rewritten for turbulent flow by introducing mixing length as

$$d = \eta^*\left(\frac{d\bar v_x}{dz}\right)^2 = \varrho l^2\left(\frac{dv_x}{dz}\right)^3 = \varrho l^2 G^3,\qquad(9.14)$$

whence

$$G = \sqrt{\frac{d}{\eta^*}} = \sqrt{\frac{D}{\eta^* V}} = \sqrt[3]{\frac{d}{\varrho l^2}} = \sqrt[3]{\frac{D}{\varrho l^2 V}}.\qquad(9.15)$$

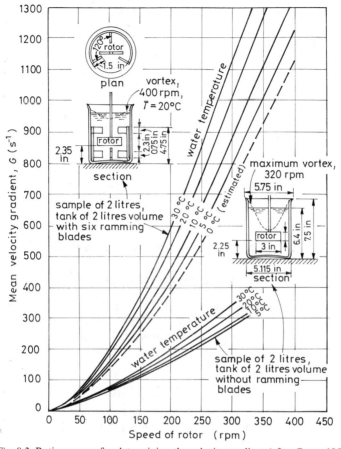

Fig. 9.2. Rating curves for determining the velocity gradient (after Camp 1955)

Mixing length l can be expressed in terms of D, G and mass M as

$$l = \sqrt{\frac{d}{\varrho G^3}} = \sqrt{\frac{D}{\varrho G^3 V}} = \sqrt{\frac{D}{MG^3}}. \qquad (9.16)$$

Mixing length l, as defined by eq (9.16), can be considered an average value — for a given structure and for given operation parameters — and can be used as a characteristic measure of turbulence. For characterizing the intensity of mixing, the energy introduced into a unit volume or unit mass, as expressed from eq (9.16) can be used in practical situations:

$$d = \varrho l^2 G^3; \qquad D = Ml^2 G^3; \qquad \frac{D}{M} = \frac{d}{\varrho} = l^2 G^3. \qquad (9.17a\text{—}c)$$

Finally, the question of measuring G will be discussed. The relevant literature generally uses eq (9.7) for calculating G. In flocculators equipped with mixers, the input energy is calculated as

$$d = \frac{2\pi g n M_n}{60\,V}, \tag{9.18}$$

where n is the speed of rotation of the mixer and M_n is the torque. Using eq (9.18) the value of G can be calculated as

$$G = \sqrt{\frac{d}{\eta}} = \sqrt{\frac{2\pi g n M_n}{60\eta\,V}}. \tag{9.19}$$

In the case of laboratory investigations, it is desirable to determine the velocity-gradient rating curves experimentally. An example for this is presented in *Fig. 9.2.*

9.3 Perikinetic coagulation

Collision of particles upon the effect of Brownian motion

In the course of perikinetic coagulation, the collision of particles — upon the effect of Brownian motion — can be considered a diffusive transport process. The relationship that describes the velocity of collision has been derived by von Smoluchowski (1917) on the basis of an electrostatical analogy, although it is essentially based on Fick's diffusion law:

$$I_{(Brown)} = \frac{dN_p}{dt} = 2\pi N_1 N_2 D_{12}(d_1 + d_2). \tag{9.20}$$

For monodispersive systems $(d_1 = d_2)$, the basic equation of perikinetic coagulation is analogous to the second-order reaction kinetic equation in the form of:

$$-\frac{dN}{dt} = 8\pi dD N^2, \tag{9.21}$$

where $I_{(Brown)}$ is the number of perikinetic collisions — upon the effect of Brownian motion — per unit volume per unit period of time; N_p is the number of perikinetic collisions per unit volume; N_1, N_2 and N are the number of particles of diameters d_1, d_2 and d, respectively, in a unit volume; D_{12} is the sum of diffusion constants for particles of sizes d_1 and d_2; if $(d_1 + d_2) = 2d$ then $D_{12} = 2D$. (The negative sign indicates that the value of N decreases with the increase of time t.) It is to be noted that in eq (9.21) d should, theoretically, be replaced by radius R of the sphere of the range of influence of the particle, which means that the $R \approx d$ approximation is made.

The conditions of validity of the above relationships are: the particles of the monodispersive system are spherical, the effect of Brownian motion only prevails, and the initial concentration of particles is appropriately high.

The basic kinetic equation

Solving differential eq (9.21), the well-known second-order reaction kinetic equation is obtained:

$$N = \frac{N_0}{1 + 4\pi dDN_0 t} = \frac{N_0}{1 + \beta t} = \frac{N_0}{1 + \dfrac{t}{T}} \tag{9.22}$$

and

$$\frac{1}{N} - \frac{1}{N_0} = k_2 t, \tag{9.23}$$

where N_0 is the initial number of particles, i.e., the initial concentration; $T = 1/4\pi dDN_0$ is the time during which the number of particles decreases to half, in the course of the coagulation process; $\beta = 1/T$; $k_2 = 4\pi dD$ the kinetic constant of the second-order process model. The factor of the efficiency of collision α_{pe} of the perikinetic coagulation can be applied as a correction multiplier of constant k_2.

α_{pe} is actually the ratio of the number of collisions resulting in coagulation to the total number of collisions, expressing the degree of effectiveness. Half-time T varies, according to practical experiences, between 2 seconds and 2 minutes. At a temperature of 298 °K $T \approx 2 \times 10^{11}/N_0$. The order of magnitude of N_0 is 10^9—10^{11} cm^{-3}.

Diffusion constant D can be calculated by the Stokes–Einstein equation as

$$D_{12} = \frac{4kT}{3\pi\eta(d_1 + d_2)}$$

or

$$D = \frac{RT}{N_A} \frac{1}{3\pi\eta d}, \tag{9.24a—b}$$

where $k = 1.38 \times 10^{-16}$ erg/°K, the Boltzmann constant; T is the temperature of the liquid in °K $(273 + °C)$; R is the gas constant, and N_A is the Avogadro number.

Comparing eqs (9.20), (9.21) and (9.24), it is found that the number of collisions I_{Brown} is independent of size d or $(d_1 + d_2)$ of the particles.

Practical conclusions

Several practical conclusions can be drawn on the basis of the above relationships. First, it can be concluded that the rate of perikinetic coagulation is a temperature-dependent variable. The effect of temperature is exercised through the variations of T and η, that define diffusion factor D. In the temperature range 0 °C—25 °C, the variation can be taken as a round 100%. It can also be concluded that flow conditions — the magnitude and distribution of flow velocities — do not play a role in the above relationships.

Nevertheless, practical experience indicates that intensive mixing, as the first phase of coagulation, is required. This is needed for the efficient mixing of flocculation agents. According to the research results of Haney (1956), intensive mixing of a few seconds duration only can be sufficient. Talman proposes 3—4 minutes of intensive mixing, while Nordell suggests 5.0—7.5 minutes. Thus mixing basins of only small volume would be required to satisfy short residence times. Intensive mixing does not disturb perikinetic coagulation, due to the reasons presented above.

9.4. Orthokinetic coagulation

The Brownian motion of coagulation acts in all directions. If the movement of particles is directed by applying certain forces, then the probability is that the number of collisions will be increased. This directing force might be gravitational, centrifugal or frictional; the resultant effect of these represents the orthokinetic stage of coagulation. Internal friction and viscosity create the velocity-gradient distribution. In this section, coagulation as induced by gravitational forces, i.e., by the settling velocity, and also by the velocity gradient (in laminar and turbulent ranges) will be discussed.

Collision of settling particles in stagnant water bodies (the effect of settling)

Let us now investigate, following the classical example, the number of collisions I_{12} of spherical particles of diameters d_1 and d_2 and settling velocities v_1 and v_2 in a unit volume of suspension during a unit period of time, assuming hydrostatic conditions. This can, obviously, be associated with the probability of attachment (linkage) of particulate and flocculate particles. On the basis of theoretical considerations it can be written that

$$I_{12(\text{stagnant})} = N_1 N_2 (\pi/4)(d_1 + d_2)^2 (v_1 - v_2),\qquad(9.25)$$

where N_1 and N_2 are the number of particles of d_1 and d_2 diameters, respectively, in a unit volume of water.

Assuming the validity of Stokes' law, it can be written that

$$(v_1 - v_2) = \frac{g(d_1^2 - d_2^2)}{18}\frac{\gamma_1 - 1}{v}$$

and by substituting this to eq (9.25), one finally obtains that

$$I_{12(\text{stagnant})} = N_1 N_2 \frac{\pi}{72} g \frac{\gamma_1 - 1}{v}(d_1 + d_2)^3 (d_1 - d_2).\qquad(9.26)$$

Conclusion: In order to increase the number of collisions, higher suspended solids concentrations (proportional to numbers N_1 and N_2), higher specific weight γ_1, the lowest possible v, higher particle sizes and the largest possible size difference $(d_1 - d_2)$, should be favoured. It was found that the size of the particle has a specially high weight. In eq (9.26), $I_{12}=0$ for $v_1=v_2$, a fairly trivial conclusion. Under the conditions mentioned above, the probability of particle collision and linkage will be increased: particles of larger dimension d_1 carry away with them particles of diameter d_2, thus accelerating sedimentation. Although the above theoretical example was derived for particulate matter, the qualitative conclusions hold for flocculent matter as well.

According to the investigations of Müller–Neuhaus (1952), the conditions

$$a_1 \geqq \sqrt[4]{40F}; \qquad a_2 \geqq \sqrt[4]{1.2F}; \qquad \frac{a_2}{a_1} \approx 0.41 \qquad (9.27a\text{—}b)$$

must be valid for particles of sizes $d_1=2a_1$ and $d_2=2a_2$ in order to allow coagulation to take place, upon the settling process. In the above equations, F can be calculated as

$$F = \frac{kT}{\pi g \varrho_1}. \qquad (9.28)$$

In practical cases $a_1 \geqq 1$ μ and $a_2 \geqq 0.41$ provide good approximations. According to Müller–Neuhaus (1952), the rate of orthokinetic coagulation in stagnant water will be optimum at approx. $d_2/d_1 = 0.8$. It eventually follows from the above considerations that this type of coagulation takes place only in polydisperse systems.

Collision of particles in laminar flow (the effect of velocity gradient)

The effects of flow conditions and mixing on flocculation can be also interpreted in terms of the number of collisions N_c and depending on the relationship between the velocity of collision and the velocity gradient. The number of orthokinetic collisions N_{ort} of water of particles 1 and 2 in a unit volume is defined by the basic equation of Smoluchowski–Camp as a function of the velocity gradient, assuming laminar flow, as

$$I_{12(\text{grad})} = \frac{dN_{or}}{dt} = \frac{1}{6} N_1 N_2 G (d_1+d_2)^3 = \frac{4}{3} N_1 N_2 g R^3. \qquad (9.29)$$

For monodisperse systems, Levics (1958) obtained the following formula:

$$I_{(\text{grad})} = \frac{4}{3} N^2 G d^3, \qquad (9.30)$$

which is in accordance with eq (9.29).

For the settling process in a stagnant water body — in the validity range of Stokes' law — variable I_{12} is defined by eq (9.25). The effect of flow and velocity gradient can be conveniently expressed in terms of the ratio of I_{12}'s in the two conditions:

$$\frac{I_{12(\text{grad})}}{I_{12(\text{stagnant})}} = \frac{12v}{\pi g} \frac{G}{(\gamma_1-1)(d_1-d_2)}. \qquad (9.31)$$

As was already mentioned, $I_{12(\text{stagnant})}$ denotes the number of collisions, per unit volume per unit time, due to the difference in settling velocities, while $I_{12(\text{grad})}$ is the number of collisions due to the flow velocity gradient.

Changes of G as functions of particle size and specific weight are shown in *Table 9.1*. The values underlined are in the domain of $10 < G\,(\text{s}^{-1}) < 75$ velocity which is favourable from the viewpoint of settling and mixing. The practical conclusions obtainable from the above Table from Fair, Geyer and Okun (1968) can be summarized as follows: mixing affects the attachment of suspended particles favourably if — according to the classification of the Table — d_1 and d_2 are large and γ_1 is small; or when d_1 and d_2 are medium and γ_1 is medium as well; or when d_1 and d_2 are small and γ_1 is high. Mixing is also inefficient when d_1, d_2 and γ_1 are equally high, or when d_1 and d_2 are medium and γ_1 is small and finally when d_1 and d_2 are small and γ_1 is medium or small. In controlling the intentsity of mixing and the extent of turbulence the above optimal range of G provides the guidelines.

It is to be noted that the effect of flow conditions on the rate of coagulation can also be characterized by ratio $I_{(\text{grad})}/I_{(\text{Brown})}$. Combining eqs (9.20) and (9.29):

$$\frac{I_{(\text{grad})}}{I_{(\text{Brown})}} = \frac{\frac{1}{6} N_1 N_2 G (d_1 + d_2)^3}{2\pi N_1 N_2 D_{12}(d_1 + d_2)} = \frac{G(d_1 + d_2)^3}{12\pi D_{12}(d_1 + d_2)} =$$

$$= \frac{G(d_1 + d_2)^2}{12\pi D_{12}} = \frac{G(d_1 + d_2)^3 \eta}{16 kT} = \frac{G R^3 \eta}{2 kT}, \tag{9.32}$$

where $D_{12}(d_1 + d_2)$ can be substituted by using eq (9.24) of Stokes–Einstein and $R = (d_1 + d_2)/2$, the effective radius, the distance between the centres of the two particles to be linked with each other (if $d_1 = d_2 = d$, then $R = d$). For monodisperse systems. Levics (1958) derived the following relationship:

$$\frac{I_{(\text{grad})}}{I_{(\text{Brown})}} = \frac{\frac{4}{3} N^2 G d^3}{8\pi D N^2 d} = \frac{G d^2}{6\pi D}, \tag{9.33}$$

Table 9.1. Variation of velocity gradient as a function of particle size and specific weight (after Fair et al. 1968)

Particle diameter (cm)		Velocity gradient G (s^{-1})				
		specific weight (g/m^3)				
d_i	d_j	high	medium	small		
		2.5 1.5	1.15	1.01	1.001	
10^{-1}	10^{-2} (high)	— —	200	20	2.0	
10^{-2}	10^{-3} (medium)	270 88	20	2.0	—	
10^{-3}	10^{-4} (small)	27 8.8	2.0	—	—	

Table 9.2. Relationship between
$I_{(grad)}/I_{(Brown)}$, the velocity gradient G and particle size d

$\dfrac{I_{(grad)}}{I_{(Brown)}}$	Velocity gradient G (s^{-1})			Limits d (μm)
	10	100	1000	
	particle size d (μm)			
10	2.8	1.0	0.5	<3
1/10	0.5	0.28	0.1	>0.1

Remark: Both effects are exercised in the range
$d = 0.1—3.0$ μm.

which is in accordance with eq. (9.32), when making the substitutions $D_{12} = 2D$ and $(d_1 + d_2) = 2d$.

In calculating with the above equation, some corresponding values are shown in *Table 9.2* for the range $10 < G$ (s^{-1}) < 1000.

Making use of eqs (9.32) and (9.33), the effects, relative to each other, of perikinetic and orthokinetic coagulation can be determined. This can be illustrated by an example from Ives and Müller (in Ives 1969). The rates of the two flocculation stages are equal when $I_{(grad)}/I_{(Brown)} = 1.0$. From this condition, $R = 2 \times 10^{-4}/G^{1/3}$ (cm) $= 2/G^{1/3}$ (micron), a simplified relationship, is obtained for a 20 °C temperature ($T = 293$ °K) and for $\eta = 0.0101$ poise. For example, for $G = 1.0$ s^{-1}, $R = 2$ μ or $d_1 = d_2 = d = 2$ μ. Consequently, it might be stated that velocity-gradient-induced coagulation is of special importance and weight if $d > 2$ μ. At $R = 20$ μ $I_{(grad)}/I_{(Brown)} = 1000$, indicating that in this range orthokinetic coagulation is round 1000 times faster than perikinetic coagulation.

Basic kinetic equations

It is also expedient to discuss the basic equation that describes the time variation of orthokinetic coagulation, similarly to the case for the perikinetic coagulation process. In doing this, the basic equation will be eq (9.29), emphasizing that only velocity-gradient-induced coagulation will be considered, while the effects of the difference in settling velocities [as described previously by eq (9.25)] will be neglected.

Let us first examined (after Hudson 1965) the simple case when $d_2 \gg d_1$ and $N_2 = $ const. Introducing $N_1 = N$ and the volumnar concentration C (the volume of silt in a unit volume of the suspension):

$$C = \frac{d_2^3 \pi}{6} N_2; \qquad d_2 = \sqrt[3]{\frac{6C}{\pi N_2}} \tag{9.34}$$

and the first-order reaction kinetic model of orthokinetic coagulation is obtained as

$$-\frac{dN}{dt} = \frac{CG}{\pi}N. \tag{9.35}$$

The substitution $dN_{or}/dt = -dN_1/dt$ can be made with the assumption that colliding particles might remain attached to each other ($\alpha_{ort} = 1$). The negative sign indicates that the value of $N_1 = N$ decreases with time t.

For the initial condition $N = N_0$ at $t = 0$, eq (9.35) can be integrated to obtain

$$N = N_0 \exp\left(-\frac{CGt}{\pi}\right). \tag{9.36}$$

This approximate solution can be interpreted (after Hudson 1965 and Ives 1969) for the case of sludge-curtain type clarifiers, where N_1 and d_1 correspond to suspended particles naturally present in the water to be treated, and N_2 and d_2 to the flocs in the flocculator. If the value of the coefficient of collision efficiency α_{ort} is less than 1.0, then C in eqs (9.35) and (9.36) should be substituted for by $\alpha_{ort}C$.

The relevant literature frequently refers to the work of Harris et al. (1966), in which the Hudson formula has been generalized, to a certain extent, in the form of

$$\frac{N_0}{N} = e^{KDGN_0t}; \qquad K = \frac{\alpha_{ort}\alpha^3 C}{\pi N_0} \tag{9.37a—b}$$

and for a cascade consisting of m units, the approximate formula is obtained as

$$\frac{N_0}{N} = \left(1 + KDGN_0\frac{t}{m}\right)^m, \tag{9.38}$$

where K is the rate coefficient of flocculation; α is a factor relating to the effective radius of flocculation and D is the particle size distribution function. For the case of $m = 1$, the relationship between flocculation efficiency N_0/N and the affecting factors G, t and N_0 is presented in *Fig. 9.3*. For the case where $m = 1$; 2 or 4, *Fig. 9.4* provides an illustrative example. It is to be noted that Soucek and Sindelar (1967) published a similar solution in the form of

$$N = N_0 \exp\left(-\frac{2}{3}K_1 K_2 \Phi CGt\right), \tag{9.39}$$

where K_1 and K_2 are experimentally quantifiable factors (e.g., K_2 is the shape coefficient of the particles) and Φ is a factor characterizing the quality of the suspension.

Collision of particles in turbulent flow (effect of the velocity gradient)

The collision of particles in turbulent flow and the related flocculation processes have been investigated by several researchers. Levics' research (1958) results in monodisperse systems resulted in the following relationship for the rate of collision

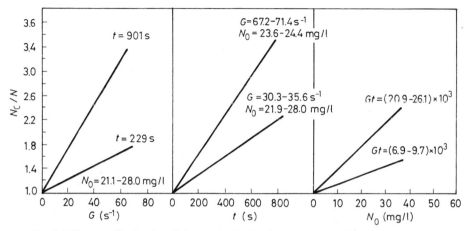

Fig. 9.3. Factors affecting the efficiency of orthokinetic coagulation (after Sontheimer 1965)

of particles in turbulent flow:

$$I_{(turb)} = \gamma \left(\frac{d}{2}\right)^3 N^2 \sqrt{\frac{W}{\eta V t}} = \gamma N^2 G \left(\frac{d}{2}\right)^3,$$ (9.40)

where the factor γ is (in this case) a multiplier factor and not the specific weight; W is the work related to energy dissipation and N is the number of particles in a unit volume.

Relationships corresponding to turbulent flow are of greater practical importance, since laminar flow occurs but rarely in treatment facilities. In this context, the

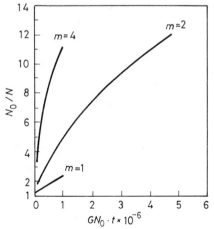

Fig. 9.4. Efficiency of orthokinetic coagulation in the case of a cascade-type flocculator (after Sontheimer 1965)

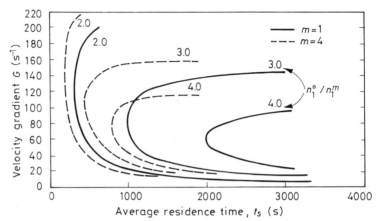

Fig. 9.5. Efficiency of orthokinetic flocculation in the case of a cascade-type flocculator (after Argaman and Kaufmann 1970)

investigations by Argaman and Kaufmann (1970) can be referred to, that aim at describing the effect of turbulence on the flocculation process. They found that the effectiveness of flocculating agents is also affected by the characteristics of turbulence, in addition to the effects of residence time, its distribution, and the input energy. The effects of the energy spectrum were determined. A figure of practical significance, based on their results, will be discussed below.

Figure 9.5 shows the relationship between velocity gradient G and residence time t_c for m series connected flocculator units. The conclusions readily obtainable from this figure are: a) for a given efficiency of the flocculator, a certain minimum detention time t_c is required; b) for a given residence time, an optimum G value can be determined; c) the splitting up of the flocculation volume (i.e., by a cascade-type design) plays an important role in respect to the hydraulic and technical efficiencies. For example, for $m=4$ units, a retention time of $t_c=800$ s is required to achieve the $N_1^0/N_1^m=4$ efficiency value, while in the case of a single unit ($m=1$), the required retention time would be 2000 s (where N_1^0 is the number of particles in the unit volume of the not yet flocculated influent, and N_1^m is the number of particles in the unit volume of water leaving the m-th cascade unit).

9.5 Superimposition of perikinetic and orthokinetic coagulation processes

In the general case perikinetic and orthokinetic flocculation take place simultaneously, both in time and space. The dominating role of either of these processes can only be decided for given specific cases. Theoretically, however, there is always superimposition of the two processes. Making use of the relationships discussed previously, the kinetic equation describing the joint effect of the two coagulation stages on the rate of flocculation can be written as

$$-\frac{dN}{dt} = k_2 N^2 + k_1 N,$$ (9.41)

where k_1 and k_2 are the first and second-order reaction rate coefficients, respectively, to be calculated [in the validity range of eqs (9.35) and (9.21)] as

$$k_1 = \frac{CG}{\pi}; \qquad k_2 = 8\pi dD = 8\frac{4kT}{3\eta}.$$ (9.42a—b)

The coefficients α_{ort} and α_{per} of the efficiency of collision can be taken into consideration as multipliers of k_1 and k_2. It can be seen that the control of hydraulic conditions could only appreciably affect the value of k_1 — that is to say that only the orthokinetic flocculation rate can be affected hydraulically.

9.6 Coagulation of particles of shapes other than spherical

In discussing the kinetics of coagulation, a spherical shape of the particles is usually assumed in order to allow a more simple mathematical discussion. In reality, however, the shape of particles differs from spherical. In this case — as is proved by experiments — the rotating motion of the particles gains significance in addition to the Brownian motion. A finding of practical importance in this respect is that the rate of coagulation of non-spherical particles (for example of bacilliform particles) is faster than that of spherical ones. In the case of aggregates formed by several anisodimensional particles, however, the effects will again approach that of spherical particles.

In the field of research into the coagulation phenomena of anisodimensional particles the colloid-physical results of Müller (in Ives 1969) should be referred to. For the effective radius R of particles of the shape of ellipsoids of rotation, Müller suggested the following relationship:

$$R = 2\ln(2a+b),$$ (9.43)

where a and b are the length of the half-axes of the ellipsoids.

9.7 Coagulation in polydisperse systems

In the above discussions, approximations assuming a uniform diameter d or two different particle sizes d_1 and d_2 were applied. In practice, however, polydisperse systems are encountered in most of the cases. The more so, since even the monodisperse systems will be turned into polydisperse systems by the coagulation proper (with the exception of coagulation due to the effects of the differences in settling velocities, that occur only in polydisperse systems) Wiegner and other authors have experimentally verified that coagulation is faster in polydisperse systems. This phenomenon is known as the "Wiegner-effect". This can be explained in an illustrative way, namely that particles of larger size actually filter out the smaller ones from the suspension — a phenomenon of special importance in the field of water and sewage treatment technology. One of the mathematical explanations is related to the higher probability of the collision of particles of differing size. As contrasted to this, the classical relationships of Smoluchowski (1917) consider identical probabilities only.

9.8 The Camp number and its modification

In Section 9.3 brief reference was made to the role of the dimensionless number $Ca = Gt_c$. Based on research results of Camp (1955), other authors have shown that the Ca number alone is not unambiguously characteristic of the process of orthokinetic coagulation. Experimental results have indicated that it is more appropriate to consider the product $C\,Ca = CGt_c$ [see also eq (9.36)]. Soucek and Sindelar (1967) refined this approach even further. Based on the theoretical considerations already referred to, these authors introduced the following criteria to characterize the coagulation process [see eq (9.39)]:

$$K_r = \Phi C\,Ca = \Phi CGt_c. \tag{9.44}$$

It can be proved that the higher the value of K_r, the faster the coagulation process. Consequently, technological efficiency can be controlled by operational modifications. For a suspension of a given quality ($\Phi = $const), eq (9.44) allows drawing the following conclusions:

a) A selected value of K_r can be maintained with different values of the parameters C, G and t_c. In other words, this means that — within certain limits — some of these parameters might be increased and the others decreased in order to maintain a nearly permanent flocculation rate.

b) It was demonstrated that in the orthokinetic range of the coagulation process, G has an optimum value, exceeding which causes the disaggregation of the flocs. This represent a technological constraint on increasing G. It is to be noted that G could be further increased together with the dosage of chemical agents (e.g., polyelectrolites) that improve the strength of the flocs. Thus, for a given C value an increase of G allows the reduction of retention time t_c, and thus the reduction of the structure volume V.

c) As opposed to the intensive mixing required for perikinetic coagulation, slow mixing is justifiable in the orthokinetic coagulation stage in order to approach the optimum value of G. Operations can also be enhanced by applying flocculation cascades. The use of cascade-type flocculation units is advantageous since the floc size — prior to settling — and thus the settling efficiency, can be increased with a gradual decrease of gradient G.

d) There are economic obstacles against increasing t_c and V, although higher residence time should be secured for lower C and G values. In the practice, the usual retention time in flocculation facilities is $t_c = 10$—50 min.

e) A higher sludge concentration C allows of decreasing gradient G, while maintaining the optimum value of CGt_c. A practical realization of this concept is the vertical throughflow type sludge-curtain clarifier, without mixing. (Chemical dosage is, nevertheless, required even in this case due to reasons mentioned above in discussing perikinetic coagulation.) C might be increased by recycling sludge.

f) The product Gt_c can be the guideline value when concentration C does not play a decisive role. Such is the case of horizontal flow clarifier, as opposed to the sludge-curtain type clarifiers. In the latter case, the setting of the optimum value of Gt_c is required. According to Ives (1969), the optimum value of Gt_c, in sludge-curtain type clarifiers, is in the range of 60—120. A value of 100 might be considered a typical guideline. For average concentrations in the order of $C = 10^{-3}$, the Ca number will be $Ca = 10^5$.

9.9 Disaggregation of flocs

As has been mentioned above, hydraulic conditions, among which this velocity gradient G, affect the flocs size, stability and disaggregation. Some experimental examples of these phenomena will be discussed in this section.

Curves shown in *Fig. 9.6* indicate the effect of the intensity of mixing on the volumetric concentration of the flocs, on the basis of measurements by Camp (1964).

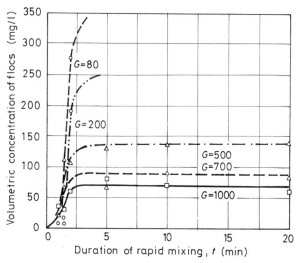

Fig. 9.6. Effect of the intensity of mixing on the volumnar concentration of flocs

15

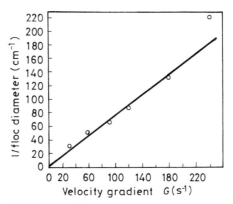

Fig. 9.7. Relationship between velocity gradient and the reciprocal of floc diameter

Figure 9.7 shows the relationship between velocity gradient G and the reciprocal of the floc diameter on the basis of experiments by Argamann and Kaufmann (1970). *Figure 9.8*, based on the experimental results of Soucek and Sindelar (1967), indicates the range of validity of kinetic equations, giving information on the disaggregation of flocs. According to this Figure, the measure of floc disaggregation is the product $G \ Re^{-1/2}$. Beyond a certain value of this quantity (the horizontal section of the curve), floc disaggregation becomes negligible. The respective experimental conditions were: alkaline medium; iron chloride as clarifying agent; $Ca = 400,000$. The Reynolds number is calculated as $Re = dv_k/v$, where d and v_k are the diameter and pheripheral velocity of the mixer rotor.

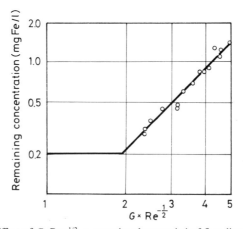

Fig. 9.8. Effect of $G. \ Re^{-1/2}$, a quantity characteristi of floc disaggregation

9.10 Main types of flocculation facilities and the principles of hydraulic design

Flocculation facilities can be grouped on the basis of hydraulic aspects as follows:

1. Rapid mixers (mixing of chemical agents, perikinetic stage).
2. Slow mixers (flocculation, the orthokinetic stage) with
   a) gravitational,
   b) mechanical, and
   c) pneumatic mixing.

Rapid mixers are mechanical equipments with a high-speed of rotation. The facilities of gravitation mixing might include: basins with diversion (baffle) walls; Venturi flumes and tubes; Parshall flumes with the establishment of hydraulic pumps; weirs, orifices, etc. Slow mechanical mixing can be performed by horizontal- or vertical-axis rotors with low speeds of rotation (sometimes equipped with built-in fixed stators). Pneumatic mixing is generally performed by blowing air into the basin (similarly as in the case with aeration basins with blown air).

In flocculation basins, the calculation of the slow mixing process — based on the velocity gradient concept — can be made in the following steps: a) calculation of input energy D; b) determination of velocity gradient G; c) $Ca = Gt_c$ or the calculation of term C Ca; and finally d) calculation of hydraulic load as Q/V.

As an example, the design steps of a gravitational flocculator, equipped with diversion walls — such as is shown in *Fig. 9.9* — will be summarized below.

a) The input energy dissipated per unit period of time is obtained as

$$D = Q\gamma h = Q\varrho gh, \tag{9.45}$$

where h is the head loss calculated as the sum of friction loss h_v [see eq (9.9)] and

$$n\frac{v_1^2}{2g} + (n-1)\frac{v_2^2}{2g} \tag{9.46}$$

(in which latter n is the number of basin compartments i.e., $(n-1)$ is the number of diversion walls). In practice $h = 15$—60 cm and $v_1 = 15$—45 cm/s.

Basin with diversion walls

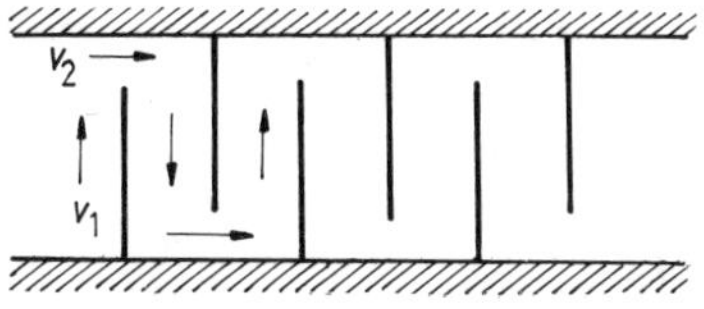

Number of basin units = n
Number of diversion walls = $n-1$

Fig. 9.9. Scheme of a gravitational flocculation basin (after Fair et al. 1968)

b) Velocity gradient G can be calculated by using the Camp–Stein formula [eq (9.7)], and the favourable range is $G = 40$—$55 \ s^{-1}$.

c) For a known retention time t_c, $Ca = Gt_c$ ($t_c = 10$—$60 \ s$); and finally

d) Hydraulic load Q/V can be calculated by eq (9.8a).

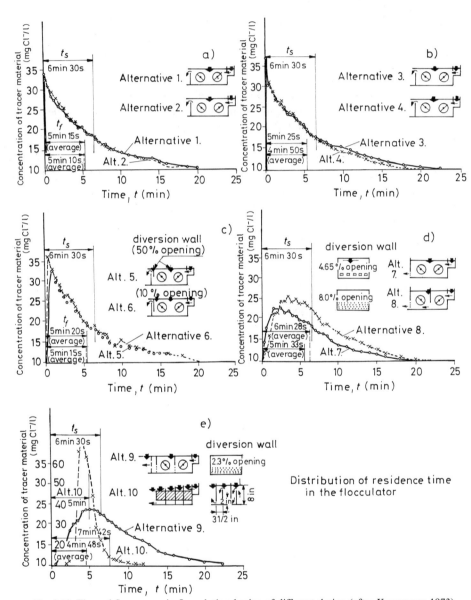

Fig. 9.10. Throughflow waves in flocculation basins of different design (after Kawamura 1973)

9.11 Distribution of residence time

An illustrative example for the hydraulic design on the basis of the distribution of residence time t_s (the throughflow wave) will be presented in conjunction with *Fig. 9.10*. Throughflow waves were measured in basins of differing design (in 10 alternatives) by determining the distribution of residence time t_c. Alternatives 1—4 were run to assess the effects of the direction of the rotation of the mixers, while alternatives 5—9 were established to investigate the effects of different perforations applied to the diversion walls. Finally, in alternative 10 the flow conditions of a cascade-type basin, without mixing, were investigated. The pheripheral speed of the mixers was 52 cm/s. The value of the ratio t_f/t_c, a quantity characteristic of the hydraulic efficiency for the ten alternatives, respectively: 0.81; 0.795; 0.835; 0.745; 0.82; 0.81; 0.85; 0.99; 1.0 and 0.96. Short-circuiting was considerable for alternatives 1—6. The direction of the rotation of mixers did not play any significant role. Comparing the Figures, it can be seen that short-circuiting was the least disturbing in the case of alternative 10, approximating the case of an ideal pipe reactor.

Due to the loss of the favourable mechanical mixing effects and to the relatively higher friction losses, however, the author did not consider this alternative the most desirable one. On the basis of operational experience, alternative 9 was considered the

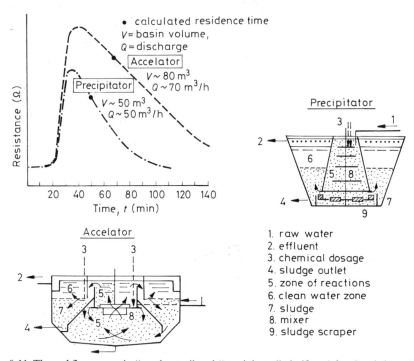

Fig. 9.11. Throughflow waves in "accelerator" and "precipitator" clarifiers (after Sontheimer 1965)

most favourable. The respective design parameters were as follows: The total area of perforations was 3% (max 5%); flow velocity in the perforated holes was max 25 cm/s (in order to eliminate splitting-up of the flocs); head loss across the perforated wall should be 0.8 cm; the suggested pheripheral speed of the mixer rotors was 36.5 cm/s; the favourable maximum velocity gradient was $G_{max} = 100$ s^{-1}; and the minimum average residence time was $t_c = 10$ minutes. The flow velocity when conveying water from the flocculator to the settling basin and to the clarifier was max 15 cm/s.

Finally, the results of the investigations by Sontheimer (1965) on the distribution of residence times in basins of "acceleretor" and "precipitator" types are presented in *Fig. 9.11.* Comparison of the respective curves indicates that the shapes of the throughflow waves are similar to each other, and thus the distribution of residence times is also similar.

9.12 Further design aids and basic data

A relationship is presented in *Fig. 9.12,* between the power consumed (HP) and the Ca number for a flocculation period of 15—40 minutes. Curves on the right-hand side of the figure refer to high-energy flocculation, when the residence time is short and velocity gradient G along with suspension concentration C are high. In this case — according to experience — flocs of high specific weight and easy settling are

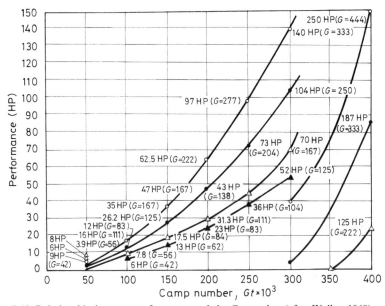

Fig. 9.12. Relationship between performance and the Ca number (after Walher 1968)

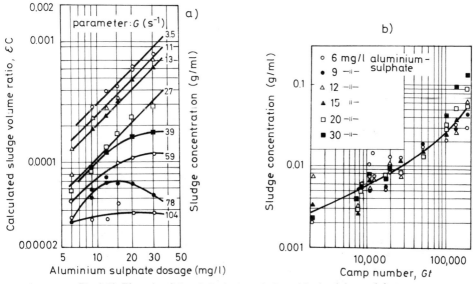

Fig. 9.13. The role of G and Gt_c in coagulating with aluminium sulphate

Table 9.3. Characteristics of flocculation when dosing with aluminium sulphate and iron chloride
(after Popp et al. 1973)

G (s^{-1})	T (°C)	Dosage								
		20			50			100		
		(W/m³)	t (min)	Gt	(W/m³)	t (min)	Gt	(W/m³)	t (min)	Gt
a) Flocculation agent: aluminium sulphate										
50	20	2.5	145.6	436 800	2.5	58.3	174 900	2.5	29.2	87 600
	10	3.25	145.6	436 800	3.25	58.3	174 900	3.25	29.2	87 600
	0	4.5	145.6	436 800	4.5	58.3	174 900	4.5	29.2	87 600
100	20	10	72.8	436 800	10	29.2	175 200	10	14.6	87 600
	10	13	72.8	436 800	13	29.2	175 200	13	14.6	87 600
	0	18	72.8	436 800	18	29.2	175 200	18	14.6	87 600
200	20	40	36.4	436 800	40	14.6	175 200	40	7.3	87 600
	10	52	36.4	436 800	52	14.6	175 200	52	7.3	87 600
	0	72	36.4	436 800	72	14.6	175 200	72	7.3	87 600
b) Flocculation agent: Iron-III chloride										
50	20	2.5	238.6	715 800	2.5	95.3	285 900	2.5	47.6	142 800
	10	3.25	238.6	715 800	3.25	95.3	285 900	3.25	47.6	142 800
	0	4.5	238.6	715 800	4.5	95.3	285 900	4.5	47.6	142 800
100	20	10	119.2	715 200	10	47.6	285 600	10	23.8	142 800
	10	13	119.2	715 200	13	47.6	285 600	13	23.8	142 800
	0	18	119.2	715 200	18	47.6	285 600	18	23.8	142 800
200	20	40	59.6	715 200	40	23.8	285 600	40	11.9	142 800
	10	52	59.6	715 200	52	23.8	285 600	52	11.9	142 800
	0	72	59.6	715 200	72	23.8	285 600	72	11.9	142 800

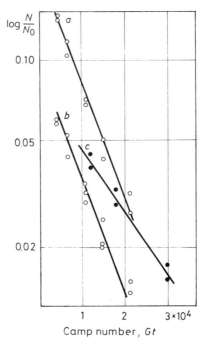

Fig. 9.14. Experimental analysis of the basic relationship of orthokinetic coagulation

formed. *Figure 9.13* shows the role of G and Gt_c when flocculating with a dosage of aluminium sulphate (after the measurement data of Hudson 1965). *Table 9.3* contains the data of flocculation with aluminium sulphate and iron chloride. It can be seen that flocculation with iron chloride needs a higher energy input.

The experimental work of Soucek and Sindelar (1967) will also be referred to. Data presented in *Fig. 9.14* verify the use of their relationship [eq (9.39)]. The experimental conditions were as follows: a) $G = 35.7$ s^{-1}; iron-chloride dosage with optimum concentration; mixing starts right after the dosage of the agent; b) same as case a) but mixing starts 20 minutes after the dosage of the agent; c) $G = 100$ s^{-1}, pH $= 9$, the chemical agent: 70 mg technical iron chloride. The straight lines connecting the points refer to an exponential relationship (due to the logaritmic scale) — providing the verification of the experiment. The favourable values of the Ca number were given by the authors as 55,000 (for aluminium sulphate) and 137,000 (for iron chloride).

On the basis of German experience, Randolf (1966) poses the following basic design data for flocculation basins: flow velocity in the basin 0.1—1.0 m/s; average residence time $t_c = 10$—20 min; friction head loss: 0.3—1.0 m. In the discharge channel between the flocculator and the settling basin, the mean flow velocity should be 0.15—0.3 m/s. Under favourable conditions, the floc size might be as high as 1.0 cm.

10 Clarifiers

Clarification is an important stage in water-treatment technology which relies essentially on the dosage of chemical agents followed by floc formation and settling. This process is applied in the field of waste-water treatment technology in treating industrial waste waters and also as the third or advanced stage of the sewage-treatment process. Discussion of the clarification process is also justified — in addition to its applications in sewage-treatment technology — by its several hydraulic implications that play a role in other waste-water treatment units (e.g., in the activated sludge treatment process).

The application of clarifiers becomes justifiable when the waste water to be treated contains large quantities of colloidal suspended particles or colloidal dissolved solids. In such cases these fine particles cannot be removed from the water by gravitational settling processes. Motion of these particles is dominated by the Brownian motion. By dosing chemical agents, the electrical charge on the surface of particles can be decreased or even eliminated, thus eliminating the repulsive forces.

10.1 Classification

According to the type of flow, clarifiers can be grouped as follows:

a) horizontal-flow clarifiers,
b) vertical-flow clarifiers,
c) transitional clarifier types (e.g., the "Uniflow" type equipment) of combined systems.

The main conditions for applying the various clarifier types are: horizontal-flow clarifier are best applicable for waters containing a relatively small quantity of fine or colloidal suspended solids; for waters of less varying water-quality characteristics; and in cases when sufficient space is available for the installation of the relatively large structures. As opposed to these conditions, vertical-flow clarifiers are best applicable to waters containing large quantities of fine or colloidal suspended solids and when water quality and load conditions fluctuate to a large extent and finally when the space available for installing the structure is restricted as far as the horizontal dimensions are concerned. It is to be noted that horizontal-flow systems are relatively less sensitive to the fluctuation of loads and temperatures — as is indicated by practical experience.

The main phases of operation are as follows:

a) chemical dosage and mixing,
b) flocculation,
c) settling

The above three operations might be implemented in separate structures or in a single combined facility. The hydraulic aspects of chemical dosage, flocculation and settling have been discussed in the foregoing chapters, and the principles formulated therein are also interpretable in the case of clarifiers. Consequently, in the subsequent sections, the hydraulic principles of the design and operation of clarifiers will be discussed with special regard to the vertical-flow systems.

10.2 Development of a suspended sludge-curtain and its stability

In vertical-flow clarifiers, various forces act on the suspended solids of the inflowing water, of which the following effects are of importance: a) downward-acting forces of gravity and b) upward-acting forces of buoyancy and friction, the dynamic forces of upward-flowing water.

The relative magnitude of these forces represent three possible situations. If gravity force $P_{down} = G$ is greater than the sum of all upward-acting forces P_{up}, then the suspended solid particles concerned are settleable. Further, if $P_{down} < P_{up}$, then the particles concerned might be washed out with the outflowing water. Finally, if $P_{down} = P_{up}$, that is in the equilibrium case, the particles remain suspended. This condition, however, can be considered a dynamic equilibrium only, as will be discussed below. In the terminology of the chemical industry, this condition is called the fluidized state.

In respect to the fate of flocs, the following consideration should also be made. When a particle is washed out from a sewage cloud and enters the zone of clarified water, then its velocity of rising will eventually — due to the law of continuity — be decreased.

When the cross-section of the clarifier structure expands in the upward direction, then the decrease of velocity is due also to geometric factors. On the other hand, the velocity of the particle also decreases in clarifiers of any design, since the velocity of water flowing through the "pores" of the sludge cloud is higher then the upward-rising velocity in the clarified zone. These effects enhance the stability of the sludge cloud.

Although under a given hydraulic condition and under identical water-quality conditions, the whole of the sludge cloud stays at a given elevation, the particles within this cloud change their location. The suspended solids content of water flowing through the sludge cloud is filtered out by the spatial "grid of flocs", thus changing the size, weight, or perhaps even the specific weight, of the individual flocs. The individual flocs thus increasingly tend to settle and move towards the lower levels of the clarifier, until finally settling takes place. Flocs thus removed from the system will be continuously and dynamically replaced by others.

The main advantage of the clarifiers that rely on the principle of developing a suspended sludge cloud is that the spatial "grid of flocs" is able to filter out — due to absorption and adhesion, etc. — even contaminants of colloidal size. This filtering effect is reflected by the term "filter-clarifiers" which is sometimes used. When the water-discharge load on the structure exceeds a certain limit, then — as is the case in fluidized beds — inhomogeneities occur within the sludge cloud, mostly taking the form of vertical "channels". Water to be treated breaks through the sludge cloud via these channels, thereby increasing the suspended solids content of the effluent. The filtration efficiency decreases and the carrying away of solids increases.

The stability of the sludge cloud can also be discussed quantitatively. One of the possible ways of thinking can be as follows: let us assume (after Ives 1968) that the clarifier is a cone standing of its tip and having a side angle (as measured from the horizontal plane) α. Then, at a distance L from the tip of the cone, the diameter D of the circular cross-section (in the horizontal plane) is $D = 2L \cot \alpha$. In a special case $D = L$ and $\alpha = 63°$, a design parameter applied in practice.

The stability equation of the sludge cloud is $v_k = w^*$, where v_k is the mean velocity of upward-flowing water and $v_k = 4Q/\pi D^2 = 4Q/\pi L^2$; and w^* is the obstructed settling velocity of the suspension. The value of this latter variable can be calculated by the Bond equation, in good harmony with experimental results, as

$$w^* = w_{St}(1 - \psi C_L^{2/3}), \tag{10.1}$$

where w_{St} is the settling velocity as obtained by Stokes' law; ψ is a shape coefficient ($\psi = 2.78$ for flocs generated by dosing with iron chloride and aluminium sulphate); and C_L is the volumetric concentration of flocs at depth L.

Substituting eq (10.1) into the equilibrium equation, concentration C_L can be expressed as

$$C_L = \left(\frac{1}{\psi} - \frac{4Q}{\psi \pi w_{St} L^2} \right)^{3/2}, \tag{10.2}$$

describing the depth-distribution of floc concentrations in the case of equilibrium.

10.3 Dissipation of energy in vertical-flow clarifiers

One of the ways of starting the hydraulic calculations is to determine the losses of energy. The loss of energy of water flowing through a clarifier — the energy dissipation — is composed of two components: the friction loss of water containing no solids (in the laminar, transitional or turbulent ranges); and the loss due to the resistance caused by the presence of flocs, the sludge cloud.

Let us characterize the loss of energy in the fluid motion by the energy dissipated during a unit period of time D (Nm/h). In a flow characterized by a water discharge Q, $Di = Q \Delta p$, where Δp is the pressure difference under the given condition.

Consider for the assumed case (after Ives 1968) a water layer of elemental depth ΔL and diameter D. The friction head loss is then obtained as:

$$h_v = 4f \frac{\Delta L}{D} \frac{v^2}{2g} = \frac{32fQ^2}{\pi^2 g D^5} \Delta L. \tag{10.3}$$

Thus the energy dissipated [see also eq (9.9)]:

$$Di = Q \Delta p = Qg\varrho h_v = \frac{32f\varrho Q^3}{\pi^2 D^5} \Delta L \tag{10.4}$$

$$d = \frac{Di}{V} = \frac{128f\varrho Q^3}{\pi^3 D^7}, \tag{10.5}$$

since

$$4f = \lambda; \qquad \Delta p = h_v \gamma; \qquad \text{further } V = \frac{D^2 \pi}{4} \Delta L$$

(in the English-language literature, resistance coefficient λ is often replaced by friction coefficient f, in accordance with the concept of Fanning, in Gould 1974).

The other component of energy dissipation is — as was mentioned above — the energy loss caused by the presence of the sludge curtain. Calculating with a suspension layer of depth ΔL, the head loss is obtained as

$$\Delta p = C_L(\varrho_1 - \varrho)g \Delta L \qquad (\text{N/m}^2), \tag{10.6}$$

since $\varrho_1 g \Delta L = \gamma_1 \Delta L$ is the hydrostatic pressure of a fluid of specific weight γ_1; $(\varrho_1 - \varrho)g \Delta L =$ the hydrostatic pressure corrected by the buoyant force, which — multiplied by concentration C_L — yields the volume of suspended solids in a unit volume of the suspension.

The energy dissipated is then obtained as

$$Di = Q \Delta p = QC_L(\varrho_1 - \varrho)g \Delta L \tag{10.7}$$

$$d = \frac{Di}{V} = \frac{4QC_L(\varrho_1 - \varrho)g}{D^2 \pi}. \tag{10.8}$$

Substituting eq (10.2) for C_L and considering $D = L$:

$$d = \frac{4Q(\varrho_1 - \varrho)g}{L^2 \pi} \left(\frac{1}{\psi} - \frac{4Q}{\psi \pi w_{St} L^2} \right)^{3/2}. \tag{10.9}$$

Ives (1968) made a comparative numerical analysis of the components of dissipated energy. He found that in the clarifier type investigated the energy dissipation due to friction losses is negligible in turbulent flow in comparison to that caused by the presence of the sludge curtain.

10.4 Velocity gradient and the Camp number

The above-described concept by Ives (1968) allows the estimation of the velocity gradient and the Camp number. Utilizing eqs (9.7) and (10.9), the velocity gradient at depth L can be expressed as

$$G_L = \sqrt{\frac{d}{\eta}} = \left[\frac{4Q(\varrho_1 - \varrho)g}{\eta L^2 \pi}\right]^{1/2}\left[\frac{1}{\psi} - \frac{4Q}{\psi \pi w_{St} L^2}\right]^{3/4}. \tag{10.10}$$

The average retention time in a layer of thickness ΔL (for the assumption of $D = L$) is:

$$t_s = \frac{\Delta V}{Q} = \frac{L^2 \pi \Delta L}{4Q}. \tag{10.11}$$

Thus, the Camp number can be calculated: $Ca = G_L t_s$. Further, the modified version of the Camp number $CaC_L = G_L t_s C_L$ can also be calculated by considering eq (10.2). Finally, for the total volume of the sludge cloud, integration yields the following formula:

$$Gt_s C = \sum_{L_a}^{L_f} G_L t_s C_L. \tag{10.12}$$

In order to allow the use of the above equations to carry out rough estimations, the following characteristic values of the variables and parameters can be taken (assuming aluminium sulphate or iron chloride dosage) as: density $\varrho_1 = 1.005$ g/cm³; dynamic viscosity of water $\eta = 0.0127$ poise (10 °C); Fanning's friction coefficient $f = 0.008$ ($\lambda = 4f = 0.032$); Bond's shape coefficient $\psi = 2.78$; $C = 0.001—0.01 = 0.1—1.0\%$; $Gt_s C = 60—120$.

10.5 Hydraulic design

In designing clarifiers hydraulically, the following two hydraulic equations are usually considered. The useful volume of the structure can be calculated by considering the average retention time:

$$V = Qt_s. \tag{10.13}$$

The second basic relationship is the continuity theorem as applied to steady-state flow:

$$Q = A_1 v_1 = A_2 v_2 = A_n v_n = \text{const.} \tag{10.14}$$

Equation 10.14 allows the calculation of the cross-section areas A_1, A_2 ... A_n required for discharging a flow Q at the prescribed velocities v_1, v_2, ... v_n.

The above equations assume a uniform velocity distribution that never exists in reality. The ideal flow pattern is modified in clarifiers by hydraulic jumps, vortices, stagnant water bodies, etc. Consequently, velocities larger and smaller than the calculated mean flow velocities occur at certain locations. These effects can be well demonstrated by coloured tracer substances. In addition to this, through-flow experiments can unambiguously indicate the distribution of residence times, and provide for the identification of the causes of dead or stagnating (sluggish) zones in the structure (short-circuiting, faulty inlet or outlet design, etc.).

As an example of hydraulic design that of the clarifier of MÉLYÉPTERV type will be presented (after a study by Bulkai 1973).

Assume a clarifier structure with the geometry of a truncated cone, turned upside down. Let the diameters of the upper and lower circular surfaces be d and $D = 2d$, respectively. Then the height of the truncated cone is $m = \sqrt{3}d/2$. The vertical cross-section of the flocculator space is a triangle of base d and height m, and the volume of this part is:

$$V_f = \frac{1}{3} \frac{d^2 \pi}{4} \frac{\sqrt{3}d}{2} = 0.226\, d^3$$

and the volume of the clarifier part of the structure is

$$V_c = \frac{4d^2 \pi}{4} \frac{2\sqrt{3}d}{2 \cdot 3} - 2 \frac{d^2 \pi}{4} \frac{\sqrt{3}d}{2 \cdot 3} = 6 V_f = 1.556\, d^3.$$

In designing clarifiers the hydraulic load T_f and the calculated average retention time t_c should equally be considered:

$$T_f = \frac{Q}{F} = \frac{Q}{\dfrac{(2d)^2 \pi}{4}} \qquad \text{and} \tag{10.15}$$

$$t_s = \frac{V}{Q} = \frac{V_f + V_d}{Q} = \frac{7 V_f}{Q}. \tag{10.16}$$

If the desired retention time of the whole structure is $t_c = 3.5$ h, then $7 V_f = 3.5\, Q$ or $Q/V_f = 2.0$ h^{-1}; $Q = 2.0\, V_f$. Substituting this latter into eq (10.15) it is obtained that

$$T_f = \frac{\sqrt{3}d}{12} \qquad \text{and} \qquad d = \frac{12 T_f}{\sqrt{3}} 7 T_f. \tag{10.17a—b}$$

Thus, if $T_f = 1.5$ m/h then $d = 10.5$ m and $D = 21$ m.

10.6 Basic design data

In the Hungarian practice of water technology, clarifiers and their related structures are designed on the basis of the following guideline values.

Inner diameter of clarifier reactors: 10—27 m; hydraulic load 180—1100 m^3/h; surface loading rate $T_f = 0.4$—0.5 mm/s; calculated average retention time $t_s = 2$—4 h; throughflow velocity at the lower part of the "bell", $v_t = 5$—10 mm/s; angle of side walls (to the horizontal) = 60°; the speed of rotation of scrapers 1—2 r.p.m.; thickness of the sludge curtain 3—5 m; thickness of the zone of treated water 1—2 m; duration of mixing chemical agents 2—3 min; flocculation time 20—30 min; speed of rotation of the flocculation equipment 6—20 r.p.m.

The importance of the Cyclofloc system should be emphasized. This system is actually a vertical flow clarifier of the sludge-curtain type. In this system, fine silt is also dosed to the water in addition to clarifying agents (in a quantity of approx. 2—5 kg/m^3), thus increasing the concentration and average density of suspended solids. This allows the increasing of the allowable load onto the structure. The characteristic surface laoding rate is 1.5—2.0 mm/s (5.4—7.2 m/h).

For design data on treating various sewages and industrial waste waters in clarifiers, handbooks on waste-water treatment technology should be consulted.

10.7 Model studies for characterizing the flow conditions in clarifiers

Experimental model studies were carried out by Ivicsics at the Water Resources Research Centre (VITUKI), Budapest, to characterize the flow conditions of settling basins. He used three approaches: a) discuss the most frequently occurring faulty flow patterns; b) made attempts at modelling the actual flow conditions in a true way and, c) the prepared design guidelines for basins with the most favourable flow conditions. A series of figures will be presented below (after Ivicsics 1968) to characterize the flow conditions developing in various clarifier (and settling) basins, when using clean water.

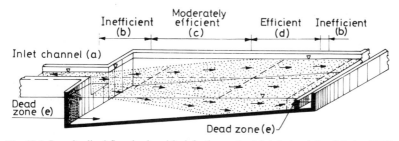

Fig. 10.1. Longitudinal-flow basin with defective water inlet design (after Ivicsics 1968)

Longitudinal flow basins

Figure 10.1 shows a longitudinal-flow settling basin with a faulty inlet design. There is a dead zone of considerable size, while flow velocities increased at other locations. These phenomena are caused, mostly, by the faulty design of the inlet structure and by the reduced spreading of the water jet caused by the same fault. *Figure 10.2* illustrates cases where the outlet channel is perpendicular to the longitudinal axis of the structure. It can be seen that the asymmetric flow distribution could not be changed, not even by installing diversion piers. A relatively favourable solution is presented in *Fig. 10.3*, where a series of orifices was applied as the inlet device. The inlet device arrangements shown in *Fig. 10.4* serve for better approaching a uniform flow distribution. Percussion disks and laths placed in front of the 3 series of inlet openings provided the desired solution in the cases *a* and *b*, respectively. The attenuating effect is further increased by the shafts and troughs installed in front of the settling space.

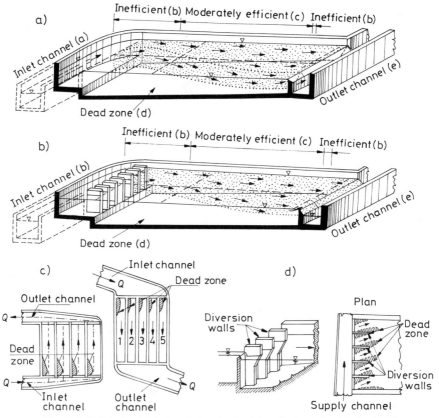

Fig. 10.2. Deficient inlet designs in longitudinal flow basins (after Ivicsics 1968)

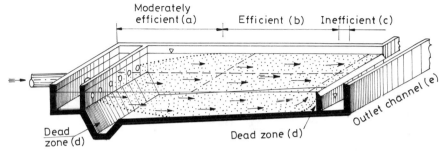

Fig. 10.3. Water inlet with perforated side wall (after Ivicsics 1968)

Radial-flow structures

There are relatively small dead zones of flow in the radial-flow settling structure shown in *Fig. 10.5a.* Hydraulic efficiency can be further increased by installing inlet elements (e.g., Stengel-heads) in several rows. An example of this is shown in *Fig. 10.5b.*

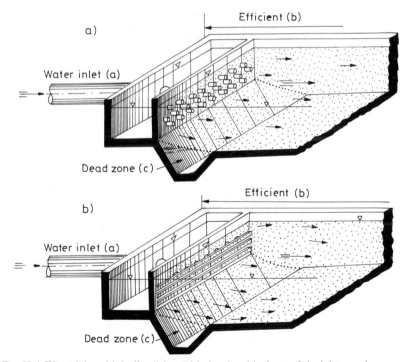

Fig. 10.4. Water inlet with buffer disks and laths placed in front of the inlet openings

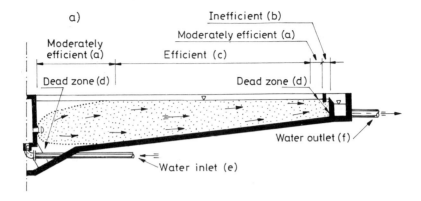

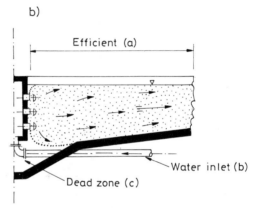

Fig. 10.5. Water inlet designs for radial-flow settling basins (after Ivicsics 1968)

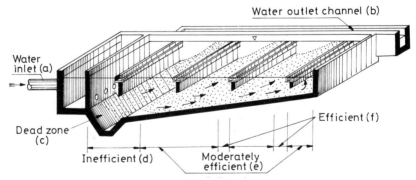

Fig. 10.6. Clarifier of the UNIFLOW type (after Ivicsis 1968)

The Uniflow type structure

Figure 10.6 illustrates the flow conditions of the Uniflow-type clarifier/settling device. In this case, water collection troughs are associated with diversion and baffle walls. It can be seen that the effect of diversion walls is unfavourable as it results in dead

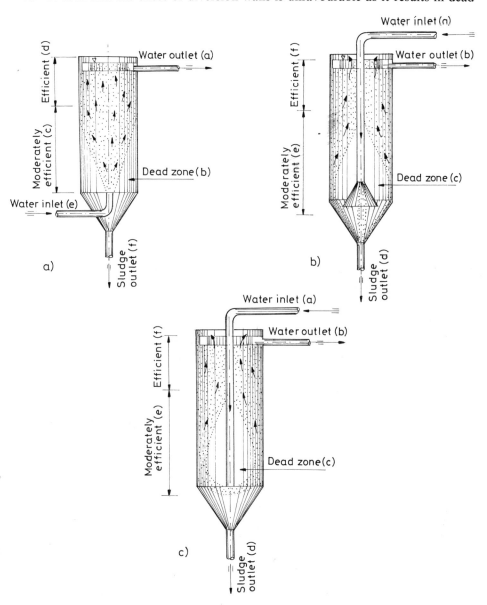

Fig. 10.7. Defective water inlet designs of vertical-flow basins

16*

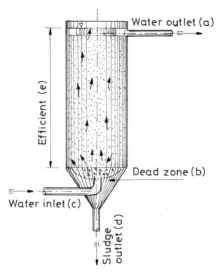

Fig. 10.8. Favourable inlet design for vertical-flow basins (after Ivicsics 1968)

or sluggish flow zones. It should be admitted, however, that water collection by multiple troughs ensures a more uniform loading pattern, but it might enhance short-circuiting to a certain extent.

Vertical flow structures

In the basin shown in *Fig. 10.7*, the utilization of space is unfavourable due to a faulty inlet design. The solutions shown in *Figs 10.7b* and *10.7c* are also objectionable from the hydraulic point of view. In the first case, a diffusor-type funnel causes dead flow zones, while in the other solution a disk, placed at the front of the opening, displaces the water jet toward the walls. Finally, the most favourable design is shown in *Fig. 10.8*, where small-size perforations made on a spherical-calotte surface assure a uniform distribution.

Corridor-type clarifiers

Figure 10.9a shows a corridor-type clarifier (half of it) with a faulty water inlet. The water jet discharging into the side unit of the basin is displaced towards one of the side walls, thus causing a sluggish flow zone on the other side. The favourable design is that shown in *Fig. 10.9b*; it allows full utilization of the available space.

Clarifiers of the MÉLYÉPTERV type

The flow conditions of the MÉLYÉPTERV clarifier, a Hungarian design made on the basis of several foreign facilities, are illustrated by *Fig. 10.10*. In the first case, the

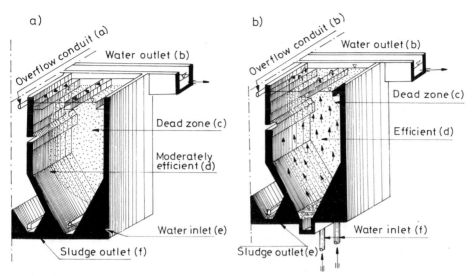

Fig. 10.9. Corridor type clarifiers with favourable and deficient inlet design (after Ivicsics 1968)

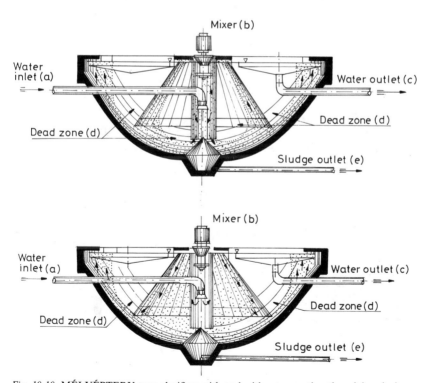

Fig. 10.10. MÉLYÉPTERV-type clarifiers with and without operating the mixing device

mixer is in operation, while in the other the mixing operation has ceased. The operation of the mixer rotor causes dead flow zones in the inner cylinder, while the downward-expanding truncated cone nappe shadows the flow.

Dual-level clarifier

The flow pattern of a clarifier of the Wittenberg type is shown in *Fig. 10.11*, with and without a mixing device. The flow pattern is more favourable when the mixer is installed in the position indicated by dashed lines. The French type of dual-level clarifier, shown in *Fig. 10.12*, is of complex design and allows the development of considerable dead zones. An inlet structure of appropriate design could significantly improve the hydraulic efficiency.

On the basis of scale-model studies, Ivicsics (1968) arrived at the following conclusions.

a) The structural design of the basins affects the velocity distribution significantly — the flow pattern is very sensitive even to small changes in the installation of the structural parts. Thus, the geometry of the structure is in close relationship with the flow pattern.

b) The inflow rate, the hydraulic load on the structure, affects the velocity distribution less significantly.

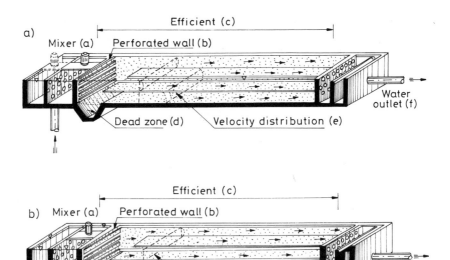

Fig. 10.11. Wittenberg-type clarifier with and without operation of a mixing device (after Ivicsics 1968)

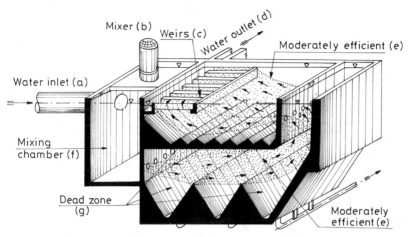

Fig. 10.12. French-type clarifier with mixed flow (after Ivicsics 1968)

c) A basic design principle is that the number and extent of changes of flow direction sould be small.

d) When instead of clean water, a fluid containing suspended solids is used in the investigations, the velocity distribution becomes more uniform.

e) The results of investigations carried out with clean water will be significantly altered when sewage water is substituted, since density and temperature conditions and their spatial and temporal variations could cause important changes in the flow domain.

f) Distinction between dead zones, as considered from the hydraulic point of view, and the technologically objectionable spatial arrangements should be made. Namely, the development of hydraulic dead zones may not only be unobjectionable but may even be expressly desirable in certain situations (for example, in the sludge zone of clarifiers).

10.8 Prototype investigations to characterize the flow pattern of clarifiers

Parallel to the above-described scale-model experiments, the associates of the author (Bulkai 1973) carried out extensive research into the hydraulics and technology of clarifiers used in Hungarian practice. The objective of these studies was to compare the operation of various existing clarifiers, and to determine the hydraulic and technological characteristics of the structure in prototype experiments. The results of these experiments have been published in Hungarian and some of the major conclusions will be presented below.

Combining hydraulic and technological considerations, the following summary statements on the operation of clarifiers can be made.

Circular, funnel-shaped vertical-flow clarifier basins proved to be favourable from both the hydraulic and the technological points of view.

Some theoretical conclusions, with due concern to reactor-technical aspects, can be drawn, thereby supplementing the practical experience. These could provide guidelines for designing other structures. The ideal "pipe-reactors" have basic importance in this context. Namely, ideal pipe reactors are characterized by a fully uniform distribution of flow velocities, which means physically that each elemental water body remains in the structure for a time t_s. Accordingly, a vertical straight line represents the through-flow wave, corresponding to the dimensionless time $t/t_s = 1$. In actual fact, this ideal case cannot be fully achieved in reality. The ideal pipe reactor can, however, be best approximated by long and narrow longitudinal flow structures. Practical experience indicates that such flow conditions well approximate those of the ideal pipe reactor, even in vertical-flow structures.

Finally, further systematic exploration of all existing Hungarian structures was recommended. Possessing an abundant amount of experimental data, the type of structure that is best applicable to certain given situations can be reliably selected. These experiments can, on the other hand, provide the possibility for improving the efficiency of existing structures and/or for expanding their loadability in a technically and economically optimum way.

11 Disinfection facilities

Disinfection facilities used in waste-water treatment technology belong to the group of chemical treatment structures. Their design and operation have several hydraulic implications, such as for example, the hydraulically appropriate dosage of chemical agents (e.g., the use of syphons), the conditions of mixing and the assurance of appropriate retention times. The elimination of sluggish flow and stagnant water bodies also falls in this category. It may be stated on the basis of the relevant literature, that only minor attention has been paid, on a global scale, to the hydraulic investigation of disinfection facilities. Design is rather restricted to general hydraulic considerations on the basis of the principles of reactor techniques, aimed at the determination of average retention times.

The work of Marske and Boyle (1973) presenting the results of comparative studies on the hydraulic properties of chlorination basins of different design and size, is one of the pioneering efforts in this respect. The hydraulics of disinfection–chlorination basins will be presented below on the basis of the above-mentioned work.

Figure 11.1 shows seven alternative basin designs along with the main design and operational parameters. The evaluation was based on Rhodamin-B tracer studies. The respective throughflow waves are shown in *Figs 11.2/a—11.2/d.*

On the basis of the prototype-scale comparative study by Marske and Boyle (1973), the following main conclusions concerning the hydraulic characterization of chlorine contact basins can be drawn.

a) The favourable flow pattern depends greatly on the conditions of the water inlet, and still more so on those of the treated-water outlet. The best water-inlet design is based on weirs that span the entire width of the basin. In this case, the unit load over the weirs can be kept rather low, thus eliminating short-circuiting. Simultaneously, the values of the parameters, characterizing the useful basin volume are favourable. As a comparative example, *Fig. 11.3* shows the hydraulic characteristics of two water-inlet designs, with a Chipoletti-weir or a sharp-edged weir, respectively. The latter design proved to be more favourable in all respects.

b) The ratio of the length to the width (L/W) of contact-chlorination basins is an important design parameter. Namely, in designing disinfection structures, the assurance of flow conditions close to those of a pipe-reactor is desirable, in order to provide

Type No.	Type of basin	Plant's mark	t_s	t_{90}/t_{10}	d	Dead space M (%)	Effective volume $(1-M)$ (%)	Plug flow P (%)	Fully mixed volume $(1-P)$ (%)	Remark:
0	ideal tube reactor			1.0	0.0	0	100	100	0	
1		A	37	2.25	0.081	10	90	72	28	
2		B	198	4.10	0.267	7	93	38	62	
3		C_1	49	2.14	0.077	15	85	71	29	
		C_2	50	2.07	0.084	0.4	99.6	62	38	
4		D_1 E	63 35	3.68 2.97	0.254 0.147	21 13	79 87	38 40	62 60	
		D_2 F	68 107	3.21 3.83	0.147 0.252	30 17	70 83	58 48	42 52	
5		G	44	2.15	0.042	20	80	75	25	wind direction is opposite to flow
		H	50	2.19	0.065	19	81	74	26	
		I	28	2.70	0.118	24	76	66	34	
6		J_1	83	1.43	0.019	42	58	96	4	Chipoletti weir: sharp-edged weir
		J_2	83	1.52	0.024	45	55	94	6	
7		K_1	50	1.29	0.007	2	98	90	10	
		K_2	47	1.36	0.011	31	69	100		

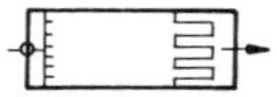

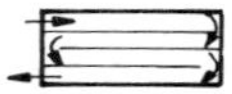

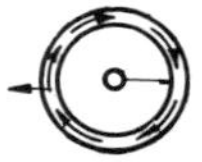

Fig. 11.1. Basin types investigated in the hydraulic experiments

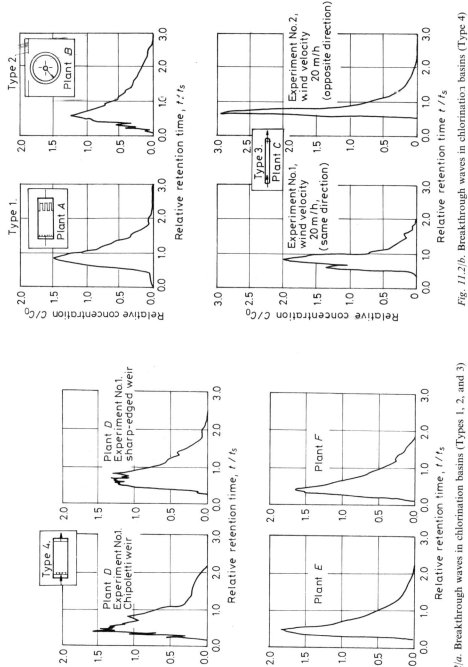

Fig. 11.2/b. Breakthrough waves in chlorination basins (Type 4)

Fig. 11.2/a. Breakthrough waves in chlorination basins (Types 1, 2, and 3)

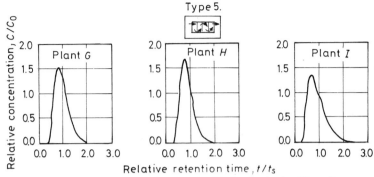

Fig. 11.2/c. Breakthrough waves in chlorination basins (Type 5)

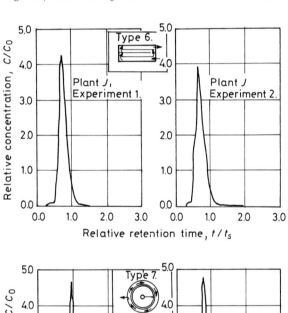

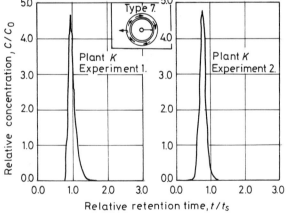

Fig. 11.2/d. Breakthrough waves in chlorination basins (Types 6 and 7)

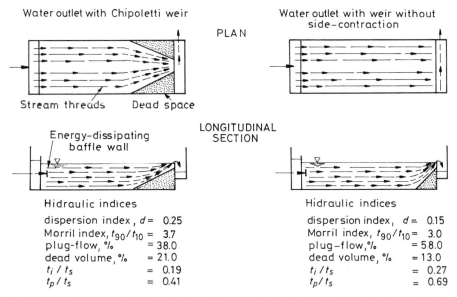

Fig. Fig. 11.3. Comparison of water-inlet facilities with Chipoletti weir and sharp-edged weir (after Marske and Boyle 1973)

a close to uniform retention time (t_c calculated as the ratio of basin volume V to the inflowing discharge Q) for each of the elemental water bodies in the basin. This is required to secure the necessary contact time for the chemical agents added. The series of experiments allowed the determination of the relationship between dispersion index d and the length-to-width ratio L/W, as shown in *Fig. 11.4*. For a ratio $L/W = 40:1$, the value of the dispersion index is approximately $d = 0.02$. It is worth

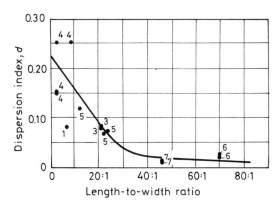

Fig. 11.4. Relationship between the dispersion index and the length-to-width ratio in chlorination basins (after Marske and Boyle 1973)

noting that above this ratio value, that is, the improvement of the efficiency is less marked.

c) Other important design aspects are related to the effect of diversion walls, as they strongly effect the length of the flow pathway in the basin. It is therefore justifiable to introduce the modified ratio L'/W', in which the actual length of the flow pathway replaces the basin length. In *Fig. 11.5* the two most widely applied practical approaches to installing longitudinal and transversal diversion walls are shown. Both the break-through waves and the hydraulic parameters indicate unambiguously that the installation of longitudinal diversion walls is more efficient. The explanation given by Marske and Boyle (1973) includes two main reasons for this: On the one hand, the extent and number of stagnant or dead flow zones is less in this latter case, whereas — on the other hand, the value of L/W ratio is more favourable, when considering identical basin volumes. In this given case, transversal diversion walls were characterized by $L/W = 20:1$ and $d = 0.085$, while the respective values for longitudinal walls are $70:1$

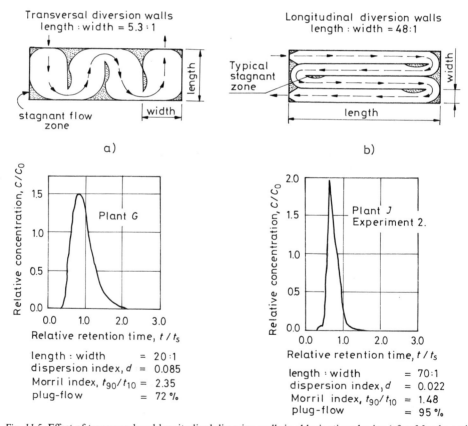

Fig. 11.5. Effect of transversal and longitudinal diversion walls in chlorinations basins (after Marske and Boyle 1973)

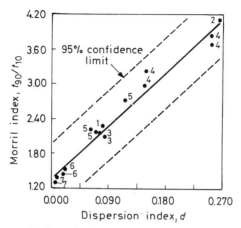

Fig. 11.6. Relationship between the dispersion index and the Morril index (after Marske and Boyle 1973)

and 0.022. Summarizing: the application of longitudinal diversion walls better approximates the conditions of *ideal pipe-reactors*.

The correlation between the Morril index t_{90}/t_{10} and the dispersion factor d was the closest among those investigated. This relationship is illustrated graphically in *Fig. 11.6*. The numbers associated with the dots in this Figure refer to various basin types investigated. Due to the close correlation, it may be stated that the Morril index characterizes the throughflow wave and the flow processes in a reliable way, in addition to dispersion index, d.

d) Wind effects were also investigated during these experiments. It was found that the velocity and direction of the wind might be of considerable effect, especially in the case of shallow basins. Wind effects can induce considerable wave motion and flow velocities in the upper water layers, which in turn may result in short-circuiting. The throughflow waves shown in *Fig. 11.2/a* clearly indicate the wind effects for the three basin types investigated. The value of the ratio t_i/t_c (t_i=initial time, t_c=calculated time), characterizing short-circuiting, was 0.036 in the case when the direction of the wind was the same as that of the flow in the basin. In the opposite case, the respective value was 0.54. The lower the value of the ratio t_i/t_c, the higher the degree of short-circuiting. The shape of the two respective throughflow waves was also different. Nevertheless, there were no significant differences in some of the hydraulic parameters, as shown in *Fig. 11.1*. The likely explanation, as given by the authors, is that in respect to the total basin volume, the effect of the wind is relatively small.

e) With respect to the geometric design of contact chlorination basins, the rectangular-shaped basin with longitudinal diversion walls (Type 6) and the ring-shaped basin around the final settling tank (Type 7) were found to be the most favourable among the seven designs investigated. According to *Fig. 11.4*, the lowest values of the dispersion index d and the highest values of the ratio L/W (exceeding 40:1 in both cases) corresponded to these two basin types. In accordance with this finding, the breakthrough waves of *Fig. 11.2/d* unambiguously indicate plug-flow, which is characteristic of pipe reactors.

f) On the basis of measurements carried out, the most desirable values of the hydraulic indices characterizing the basin geometry can also be given. Considering 40:1 for the ratio L/W, the following index values can be recommended:

dispersion index: $d = 0.02$
Morril index $= t_{90}/t_{10} = 1.55 \pm 0.62$
$t_i/t_c \quad = 0.42 \pm 0.29$
$t_m/t_c \quad = 0.81 \pm 0.23$
$t_{50}/t_c \quad = 0.83 \pm 0.22$
plug-flow volume, $\% = 87.4 \pm 20.6$
effective volume, $\% = 76.6 \pm 27.2.$

g) Hydraulic indices are interrelated, a well-known fact: in the given case, the following relationship was derived experimentally:

$$y = (A + Bx) \pm C,$$

where y is a certain dimensionless hydraulic index and the dependent variable x is the dispersion coefficient d. A and B are experimental coefficients and C is the 95% confidence interval. An actual example of this can be given as follows:

$$\frac{t_k}{t_s} = (0.44 - 1.24d) \pm 0.29; \ R = 0.44$$

$$\frac{t_m}{t_s} = (0.84 - 1.38d) \pm 0.23; \ R = 0.75$$

$$\frac{t_{50}}{t_s} = (0.85 - 0.49d) \pm 0.22; \ R = 0.38$$

$$\frac{t_{90}}{t_{10}} = (1.35 + 10.24d) \pm 0.62; \ R = 0.95.$$

h) The characteristic domains of flow could also be determined, on the basis of the method of Wolf and Resnik (1963). For example, the magnitude of dead zones (M), the effective volume ($1 - M$), the volume of plug flow (P) and the volume of full mixing ($1 - P$) can be calculated. The respective data are shown in *Figure 11.1*. The same Figure gives the characteristic hydraulic parameter values for the breakthrough waves.

III. Biological treatment

12 Aerated activated sludge systems

Aeration basins are the basic structures of activated sludge waste-water treatment systems. At a given organic load, their volumes are indirectly determined by the treatment efficiency to be achieved. The shape of the aeration tanks should be designed by taking into consideration the aeration equipment to be used, as well as those aspects necessary for assuring hydraulically favourable flow conditions. During the past few decades, significant results have been achieved in the investigation of the fluid mechanics of aeration basins.

In Hungary, aspects of the most suitable geometric/hydraulic designs were investigated at the Water Resources Research Centre (VITUKI), Budapest, and at the University of Technology of Budapest. The relationship between hydraulic properties and oxygen transfer processes for aeration basins of various types was also analyzed at the same institutions. In this respect, some of the earlier works of the present author should be consulted for further details (Horváth 1966a and 1970).

12.1 Hydraulic characteristics of aeration basins

To illustrate the fluid-mechanical properties of aeration systems, some (Hungarian and foreign) examples will be presented in this section. In *Fig. 12.1* the measurement results of Pató (1961) (in the form of isotachs) are presented for the Kessener-type aeration basin of the waste-water treatment plant at Budakeszi, Hungary. A favourable cross-sectional design is shown in *Fig. 12.2*, for an aeration system of the horizontal-axis type, based on the research results of the present author (Horváth 1970). To characterize the hydraulic properties of the INKA-type aeration system, *Fig. 12.3* shows the results of a scale-model study of the basin of the waste-water treatment plant at Hatvan, Hungary.

Among foreign examples, the flow-velocity distribution study by Möller (1963), carried out in an aeration basin with deep air injection, (Oberhausen-Holten, Germany) is mentioned first (*Figs 12.4/a—12.4/c*). Further experimental results of a similar type are presented in *Figs 12.4/d—12.4/g*. The results of hydraulic measurements carried out with aeration systems of the Vortair and Dorr–Oliver types by

a)   b)

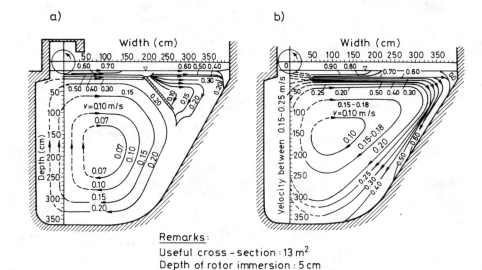

Remarks:
Useful cross-section : 13 m²
Depth of rotor immersion : 5 cm

Fig. 12.1. Flow conditions in the Kessener type aeration basin of the waste-water treatment plant of Budakeszi (Hungary) with and without baffles

a) Flow conditions in the activated sludge aeration basin of the waste-water treatment plant of Budakeszi.

b) Flow conditions in the activated sludge aeration basin of the waste-water treatment plant of Budakeszi, without baffles

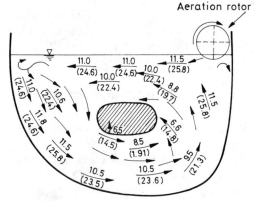

Remarks:

Depth of rotor immersion : in the model : 12 mm
                           in the prototype : 6 cm
Numbers on arrows indicate flow velocities (cm/s)

$$\left(\frac{\overleftarrow{11.0} - model}{24.6 - prototype} \right)$$

The model is based on Froude's law ($\lambda = 5$)

Fig. 12.2. Flow pattern in the scale model of a well-designed aeration basin (the geometry was based on klotoid curves)

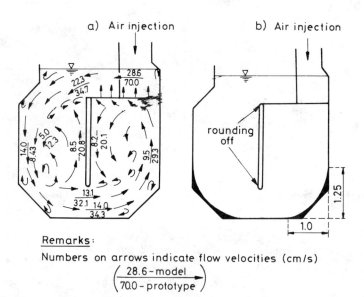

Remarks:

Numbers on arrows indicate flow velocities (cm/s)

$$\left(\dfrac{28.6 - \text{model}}{70.0 - \text{prototype}} \right)$$

Fig. 12.3. Flow pattern in the scale model of an INKA-type aeration basin, and a proposed design for archieving improved flow conditions

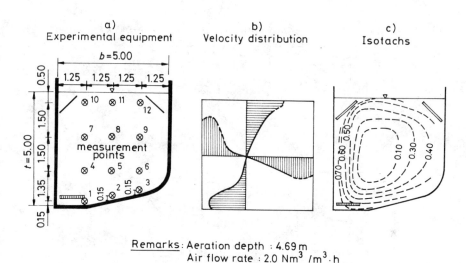

Remarks: Aeration depth : 4.69 m
Air flow rate : 2.0 Nm^3 /m^3·h
Velocities (m/s)
Fluid : tap water

Fig. 12.4/a–c. Flow conditions in an aeration basin with deep air injection (after Möller 1963)

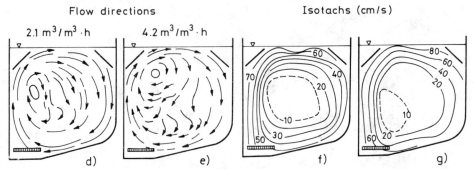

Fig. 12.4/d–g. Flow conditions in aeration basins with deep air injection (after Knop, Bischofsberger and Stalmann 1964)

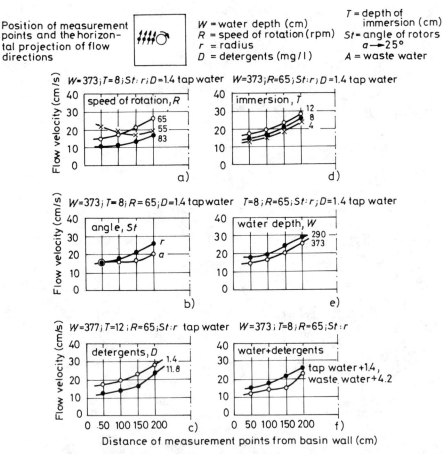

Fig. 12.5. Flow conditions in an aeration basin equipped with Vortair rotors (after Knop, Bischofsberger and Stalmann 1964)

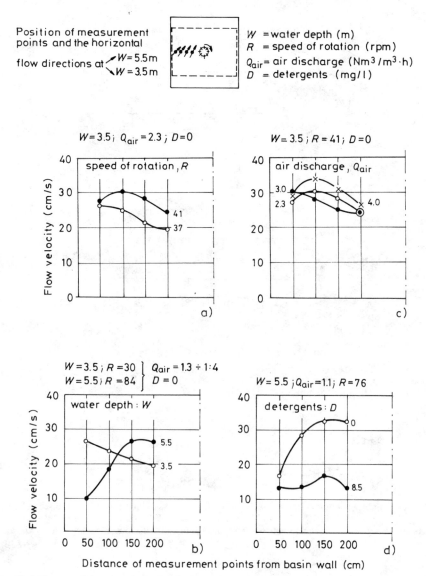

Position of measurement
points and the horizontal

flow directions at $<\begin{matrix} W=5.5\text{m} \\ W=3.5\text{m} \end{matrix}$

W = water depth (m)
R = speed of rotation (rpm)
Q_{air} = air discharge (Nm³/m³·h)
D = detergents (mg/l)

$W = 3.5$; $Q_{air} = 2.3$; $D = 0$

speed of rotation, R

Flow velocity (cm/s)

41
37

a)

$W = 3.5$; $R = 41$; $D = 0$

air discharge, Q_{air}

3.0
2.3
4.0

c)

$W = 3.5$; $R = 30$ ⎱ $Q_{air} = 1.3 \div 1.4$
$W = 5.5$; $R = 84$ ⎰ $D = 0$

water depth: W

Flow velocity (cm/s)

5.5
3.5

b)

$W = 5.5$; $Q_{air} = 1.1$; $R = 76$

detergents: D

0

8.5

d)

0   50   100   150   200

Distance of measurement points from basin wall (cm)

Fig. 12.6. Flow conditions in an aeration basin equipped with Dorr—Oliver rotors (after Knop, Bischofsberger and Stalmann 1964)

researchers of the Emschergenossenschaft are shown in *Fig. 12.5* and *12.6a*, respectively. The flow-velocity distribution formed in a shallow aeration basin of trapezoidal cross-section is illustrated by *Fig. 12.7*. Data from measurements on fine-bubble
air-diffusion systems is presented in *Fig. 12.8*. A comparison of the flow-velocity

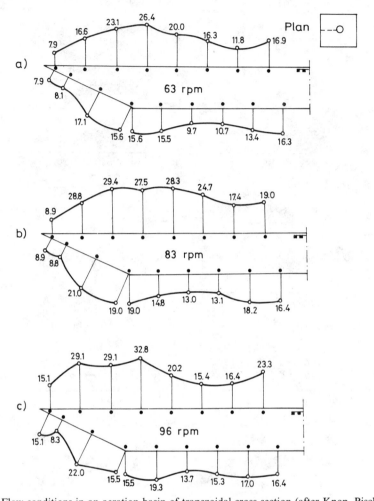

Fig. 12.7. Flow conditions in an aeration basin of trapezoidal cross-section (after Knop, Bischofsberger and Stalmann 1964)

Measurement points are at 20 cm below water surface and 15 cm above bottom. Flow velocities are in cm/s

distributions for systems with fine-, medium-, and rough-bubble air-diffusion is facilitated by *Fig. 12.9.*

As a final example, *Fig. 12.10* adopted from a study by von der Emde (1964) is presented, in which bottom-flow velocities measured in different aeration basins are plotted as a function of the power input.

Following the original idea of Kalbskopf (1966b), the characteristic hydraulic parameter in this case is specified as the spatial hydraulic radius R_r (the ratio of the wetted volume to the wetted surface area).

Direction and flow velocity
(fine – bubble aeration)
Air discharge : 4.46 Nm^3/m^3 ·h
Air discharge of single diffusor : 11.8 Nm^3/h
Number of diffusors : 52

Direction and flow velocity
(fine –bubble aeration)
Air discharge : 4.85 Nm^3/m^3 ·h
Air discharge of single diffusor : 12.8 Nm^3/h
Number of diffusors : 52

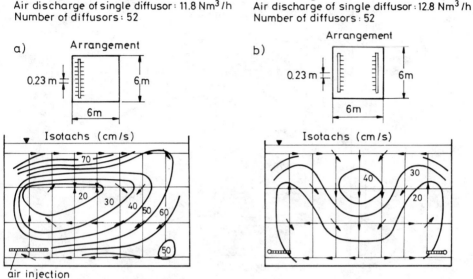

Direction and flow velocity
(fine – bubble aeration)
Air discharge : 4.45 Nm^3/m^3·h
Air discharge of single diffusor: 11.7 Nm^3/h
Number of diffusors : 52

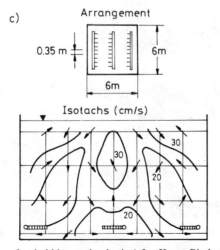

Fig. 12.8. Flow conditions in a fine-bubble aeration basin (after Knop, Bischofsberger and Stalmann 1964)

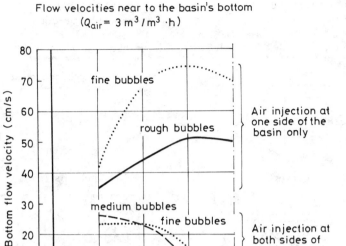

Flow velocities near to the basin's bottom
$(Q_{air}= 3\ m^3/m^3 \cdot h)$

Fig. 12.9. Comparison of flow conditions in fine-, medium- and rough-bubble aeration basins (after Knop, Bischofsberger and Stalmann 1964)

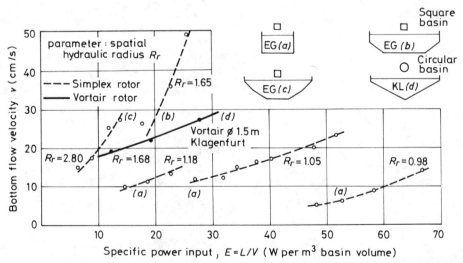

Fig. 12.10. Relationship between power consumption and bottom flow velocity in basins with vertical aeration rotors
EG = Emschergenossenschaft data (after Kalbskopf); KL = Klagenfurt data (after von der Emde)

12.2 Pumping capacity of aeration systems

The pumping capacity of aerating equipment is measured in terms of the flow rate induced by the operation of the equipment. On the basis of the method developed by Kalinske (1970) this is calculated by relying on the determination of the rate of oxygen transfer.

The total oxygen uptake capacity is calculated as:

$$Q_t = V \cdot K_L a(C_s - C_t) \quad (\text{kg O}_2/\text{h}) \qquad (12.1a)$$

or when $C_t = 0$;

$$Q_t = V \cdot OC = VK_L aC_s = VK_L \frac{F}{V} C_s, \qquad (12.1b)$$

where V is the useful volume of the aerating basin (m³); F is the interfacial area between gas and liquid, i.e., the surface of diffusion (m²); OC is the oxygenation capacity (g O_2/m^3h); K_L is the rate coefficient of mass transfer (m/h); $K_L a$ is the extended rate coefficient of mass transfer (h⁻¹); O_t is the total capacity of oxygen uptake (kg/h); C_t is the DO concentration at time t (kg/m³); C_s is the saturation DO concentration (kg/m³).

O_t can also be calculated on the basis of the rate of flow Q in the vicinity of the aeration device (the pumping capacity) as

$$O_t = Q(C_2 - C_1), \qquad (12.2a)$$

where C_1 and C_2 are the DO concentrations of the inflowing and outflowing water at the aeration device, respectively. According to of Kalinske's experimental results it is desirable to apply a correction coefficient K in the mass-balance eq (12.2a), which will account for the non-ideality of the actual conditions.

Assuming that $C_2 = C_s$:

$$Q_t = KQ(C_s - C_1). \qquad (12.2b)$$

Experimental experience indicates that the value of the coefficient K depends on the Froude number Fr ($\text{Fr} = v_k / \sqrt{gh_r}$, where v_k is the pheripheral velocity of the rotor):

$$K = m \, \text{Fr}^n, \qquad (12.3)$$

where m and n are experimental constants (for example, in the case of steeply inclined chutes and hydraulic jumps, $n = 1.0—1.4$). Combining eqs (12.1b) and (12.2b), the pumping capacity can be obtained as

$$Q = \frac{O_t}{K(C_s - C_1)} = \frac{VK_L aC_s}{K(C_s - C_1)} = \frac{VOC}{K(C_s - C_1)}. \qquad (12.4)$$

The relationship between pumping capacity and power consumption was recently analysed by the present author, taking the proportions of energy spent on oxygen transfer and on inducing flow also into consideration (Horváth 1984).

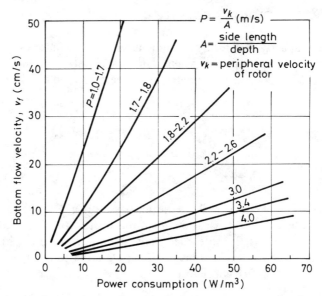

Fig. 12.11. Relationship between bottom-flow velocity and power consumption for an aeration basin equipped with BSK rotor (norm AMC. Ag)

The determination of the rate of flow in aeration basins can be done by calculating the flow across well-defined surfaces (such as the area under the separation wall in the INKA system). Muszkalay (in Horváth 1984) carried out further measurements for the quantification of flow conditions in INKA type aeration basins.

Figure 12.11 shows typical results for the pumping capacity of an aeration device. The curves demonstrate the relationship between bottom-flow velocity (a measure of the pumping capacity), and the specific power comsumption for a basin aerated with BSK rotors. The Figure also demonstrates the role of the basin geometry. (It is to be

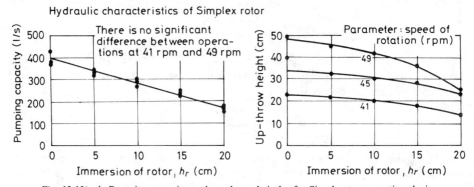

Fig. 12.12/a–b. Pumping capacity and up-throw heigth of a Simplex type aeration device

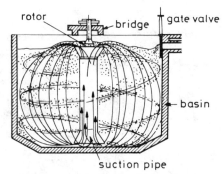

Fig. 12.12/c. Schematic diagram of the flow pattern in Simplex type aeration basin

noted that the bottom-flow velocity was measured at a distance of 1.0 m from the basin wall.)

The curves of *Fig. 12.12/a—b* were determined by the researchers of the Emscher-genossenschaft in course of the hydraulic evaluation of simplex aeration basins. The flow pattern of this arrangement is shown, schematically, in *Fig. 12.12/c.*

12.3 The effect of surface-active substances (surfactants)

Waste-waters sometimes contain chemical compounds which substantially affect the flow conditions.

These effects are mostly due to changes in density, specific weight and surface tension. Among such substances, surface-active materials (detergents) play the dominant role, significantly affecting the surface tension of flowing liquid, thus enhancing the formation of bubble-rich zones.

Kalbskopf (1966b) conducted detailed investigations in air-injected aeration basins into the changes in flow pattern resulting from the addition of surface-active substances. A set of curves, summarizing his results, is shown in *Fig. 12.13*. It can be seen from this Figure that significant effects on the bottom-flow velocity occur mostly in the range of $C < 10$ mg/l. On the other hand, the decrease of flow velocity will be less marked in systems with deep air injection.

The unfavourable effects of surface-active materials are also well demonstrable in Kessener-type aeration basins. These effects are illustrated in *Fig. 12.14*, based on the work of von der Emde (1964). The distribution of the *DO* concentration indicates that in case *a*) significant velocities are present only in the upper third of the basin. Below this, there is a dead zone of little or no flow, which induces the settling of sludge particles. It can be demonstrated that an independent flow zone is formed by the upper, bubble-rich part. Baffles (producing pressurizing and depressurizing effects) can be used, as indicated by *Fig. 12.14b*, to make the velocity-, and thus the *DO*

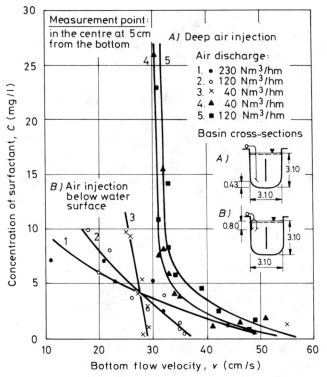

Fig. 12.13. Variation of bottom-flow velocities with the concentration of detergents in an aeration basin
with air injection (after Kalbskopf 1966)

profiles more uniform. It is to be noted that similar results have also been published
in the Hungarian literature, on the basis of experimental measurements carried out in
VITUKI (Horváth 1984).

12.4 Velocity gradients and energy dissipation

The work of Fair et al. (1968), utilized an energy-dissipation approach to the
problem of biological flocculation in activated sludge aeration basins. On the basis of
technological investigations (as shown in *Fig. 12.15*) it was demonstrated that the
treatment efficiency of activated-sludge systems can be improved by increasing the
intensity of turbulence. In their classic work Camp and Stein (1943) gave the following
equations for calculating the energy input and the average velocity gradient in
air-injected aeration basins:

$$D = \varrho g h q_{\mathrm{air}} Q_v \qquad (12.5)$$

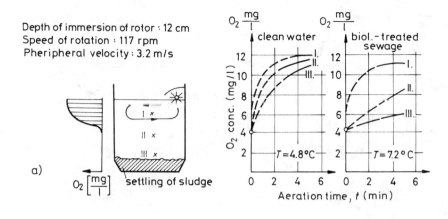

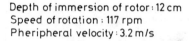

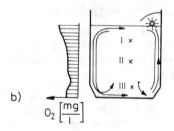

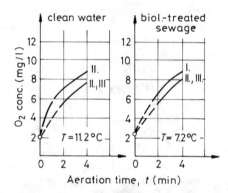

Fig. 12.14. Effects of detergents on the flow conditions and distribution of the oxygen concentration of Kessener type aeration basins

$$G = \sqrt{\frac{D}{\eta V}} = \sqrt{\frac{\varrho g h q_{air} Q_v}{\eta t_s Q_v}} = \sqrt{\frac{\varrho h q_{air}}{v t_c}}, \qquad (12.6)$$

where h is the depth of water in the basin (m); q_{air} is the air input per unit volume of sewage, (m^3 air/m^3 sewage); $t_c = V/Q_v$, the calculated average retention time (h).

It should be noted that the useful input energy must be distinguished from the total value of energy dissipation, since only a certain percentage p of the input energy will be utilized. This fact can be taken into consideration by replacing D in eq (12.6) by pD.

Research results by Zahradka (1967), Goda (in Horváth 1984), Rincke and Möller (1966) and others also justify the use of velocity gradients in evaluating biological waste-water treatment systems.

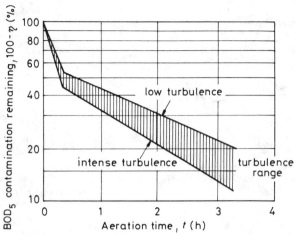

Fig. 12.15. Effect of turbulence on the relationship between treatment efficiency and retention time (after Fair et al. 1968)

12.5 Flocculation and the floc size of activated sludge

There are several theories available to explain the formation of activated sludge. McKinney (1962), Hartmann and Laubenberger (1968) analyzed the mechanism of floc formation, taking the role of hydraulic conditions also into consideration.

The formation of activated sludge flocs is, essentially, the result of coagulation and flocculation processes. In respect of the order of magnitude, there are 2-3 10^9 bacteria present in 1.0 cm^3 of sludge volume. Their size is approximately 5×10^{-5} cm. On the basis of the theory of Smoluchowski (1917), the extent of the role of perikinetic coagulation, stemming from Brownian motion, can be estimated, as an effect supplementing that of the orthokinetic coagulation defined by the flow conditions. The ratio of the probability of collision of particles has been determined by Laubenberger (1970) (*Table 12.1*) for various particle diameters and for three different velocity gradients. (The Boltzmann constant $k = 1.38 \times 10^{-16}$ erg/$°K$; $T = 300$ $°K$, $\eta = 1 \times 10^{-2}$ poise.) As indicated in the Table, the effect of perikinetic coagulation is negligible in the range of $d > 5$ μm (in this range, the values given in the Table are higher by orders of magnitude than the unit. According to the measurement results of Mueller et al. (1966), the size of activated sludge flocs in laminar flow is generally greater than 5 μm.

12.6 Size of suspended particles and stresses thereon

In addition to the forces that enhance floc formation there are eventually others that hinder these processes, causing the disaggregation of flocs. These latter forces are largely determined by the flow conditions. An equilibrium condition might also be

Table 12.1. Evaluation of the role of perikinetic and orthokinetic
coagulation in laminar flow
(after Laubenberger 1970)

Particle diameter (μ)	$I_{(grad)}/I_{(Brown)}$		
	Average velocity gradient		
	$G = 5$ (s^{-1})	$G = 20$ (s^{-1})	$G = 50$ (s^{-1})
0.5	0.075	0.302	0.755
1	0.604	2.415	6.039
5	75.483	301.932	754.831
10	603.865	2415.459	6038.647
50	75483.092	301932.367	754830.918
100	603864.734	2415458.937	6038647.343

established, when the rate of floc formation is equal to that of floc disaggregation.

Let us now consider (after the research results of Rumpf and Raasch 1962) the forces causing the disaggregation of flocs in laminar flow. Direct forces and shear forces act on the particles, causing forward and rotating motion.

Considering spherical and cylindrical particles as the ideal case, the stresses developing on the surface of a particle can be approximated by the following equations:
Shear stress:

$$\tau = 2.5\eta G \text{ (sphere)} \tag{12.7a}$$

$$\tau = 2.0\eta G \text{ (cylinder),} \tag{12.7b}$$

additional stresses due to the centrifugal force (in the case of rotating motion):

$$\sigma = 0.4\varrho\,\omega^2 r^2 = 0.1 G^2 r^2 \text{ (sphere)} \tag{12.8a}$$

$$= 0.41\varrho\,\omega^2 r^2 = 0.102 G^2 r^2 \text{ (cylinder),} \tag{12.8b}$$

where $r = d/2$ is the radius of a spherical particle; ω is the angular velocity of particle rotation; ϱ is the density of the particle.

The conditions of validity for the above equations are: laminar flow; mass-forces are negligible in comparison to surficial forces; and Hooke's law pervails in respect to the material of the particle. In order to characterize the above parameters numerically, Fig. 12.16 is presented (after Rumpf and Raasch 1962).

It is indicated by the above analysis that velocity gradient G plays a role not only in the coagulation process, but also in forming the stresses that cause the disaggregation of particles. Laboratory experiments indicate that flocs of higher strength are formed at higher G values, if all other conditions remain the same; in contrast, large flocs of loose structure are formed at lower G values. Consequently, higher G values result in the establishment of larger specific suspended-solid surfaces. This is illustrated by Fig. 12.17. In this case it is also seen that for $G = $ const the specific surface area decreases with increasing sludge concentration. This is mainly due to the higher probability of collision of particles at higher concentrations of solids.

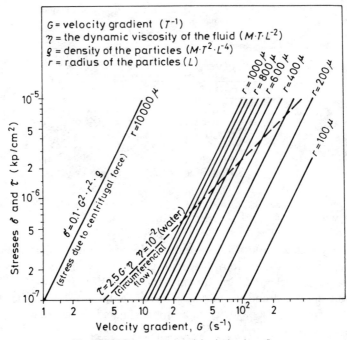

Fig. 12.16. Stresses on particles in laminar flow

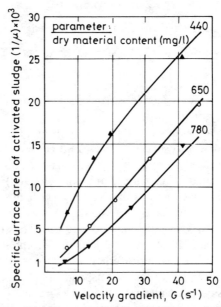

Fig. 12.17. Variation of the specific surface area of activated sludge as a function of the velocity gradient and the sludge concentration (after Laubenberger 1970)

Laubenberger (1970) summarizes the effects of hydraulic conditions on the particle size, and on the specific surface area of flocs as follows: with increasing flow velocities, i.e., at higher velocity gradients, and at given sludge concentrations, the variation of floc size decreases which means that the whole sludge volume takes part in the treatment process in a more uniform way. Measurement data indicate that under equilibrium conditions the distribution of floc size is lognormal. In the transitional range from laminar to turbulent flow, there is a step-like increase of the specific surface area of flocs. A higher specific surface area in turn facilitates mass transfer and exchange processes (oxygen and nutrient exchange processes).

According to the results of several researchers, a smaller floc size does not affect the efficiency of settling basins adversely (although there are also contradictory opinions in this respect).

Finally, it is to be noted, that the experiments by Laubenberger were carried out in a rectangular glass tube of 6×1 cm cross-sectional area.

12.7 Oxygen supply and flow conditions

12.7.1 Bubble motion

The rising velocity of bubbles depends on their size. The shape of moving bubbles is defined by the buoyant forces, the drag and the surface stresses, and their relative proportions. Surface tension tends to maintain a spherical shape and this defines the minimum surface area of the bubble. Drag resistance, on the other hand, tends to flatten out the bubbles. According to Levics (1958), deformation starts at a particle size of 0.01—0.1 cm. While spherical bubbles rise along a straight line, lens-shaped flatter bubbles move also transversally in the direction of the least resistance, resulting in a helical pathway of rising. A comparison of the measurement results of several other authors and our own experiments for the rising of bubbles in stagnant water is shown in *Fig. 12.18*.

Some analytically derived relationships concerning rising velocity of bubbles of different sizes will be presented below. For the movement of small bubbles in stagnant water, Levics obtained the following formula:

$$w = -\frac{g r_B^2}{3 v}; \qquad \frac{\mathrm{Fr}}{\mathrm{Re}} = -\frac{1}{3}, \tag{12.9a}$$

with the assumption of $\mathrm{Re} \ll 1$. The length in the Reynolds formula, in this case, is the radius of the bubble, while the characteristic velocity is that of the rising bubble. The negative sign refers to the bubble's velocity which is opposite to the force of gravity. If Re varies in the range of 50—800 then

$$w = -\frac{g r_B^2}{9 v}; \qquad \frac{\mathrm{Fr}}{\mathrm{Re}} = -\frac{1}{9}. \tag{12.9b}$$

274 III. Biological treatment

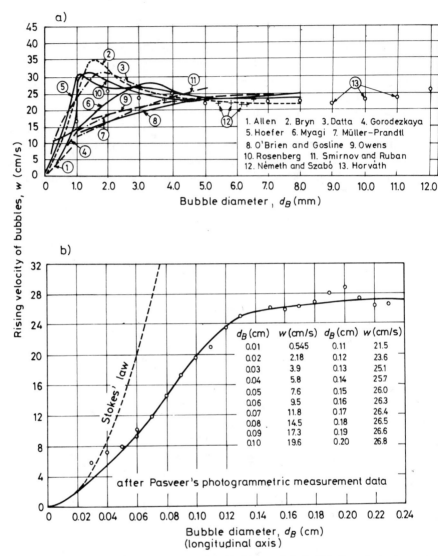

Fig. 12.18. Relationship between the rising velocity of bubbles and the bubble diameter in stagnant clean water (Horváth 1984)

In obtaining the above results it was assumed that the bubbles are of spherical shape and the water is free of surface-active substances. This latter assumption is important since their presence in the water restricts the velocity of movement of other substances (gas or solids).

In the case of bubbles larger than a radius of 0.1 cm, i.e., in the case of Re > 800,

Levics suggests the following relationship:

$$w \approx \frac{2}{3} \left(\frac{4\sigma^2 g}{3\alpha^* \varrho \eta} \right)^{1/5},$$

(12.10)

where σ is the specific surface stress and α^* is an experimental constant characterizing the motion of large bubbles.

This suggests that the velocity of rising of deformed bubbles is independent of their size. Experimental results by several researchers, in accordance with our own experiments, indicate that the limit of the validity of this relationship is $r_B = 0.25$ cm, rather than $r_B = 0.1$ cm. Although there is a further increase of velocity when $r_B > > 0.25$ cm, in the lower ranges the relationship between velocity and bubble diameter can be considered as a horizontal line. In practice, in the case of aeration basins with blown air, the bulk of the air volume is present in bubbles larger than 2.5 mm in diameter.

In *Table 12.2* a relationship is presented in dimensionless and dimensioned forms (after the book by Grassmann 1961) for calculating the velocity of rising air bubbles, and the respective drag coefficient. Stokes' law well approximates the processes in the range Re < 2. For practical purposes, the simplified relationships (presented in *Table 12.3* after Hörler 1964) can be recommended. Finally, the data of *Table 12.4* suggest that the pore or perforation size of the air-blower device cannot be considered identical with the size of the bubbles generated. The latter can be larger by as much as 8—10 times than the former.

Table 12.2. Rising velocity of bubbles in water and the drag coefficient

Characteristic ranges			Drag coefficient	Rising velocity, w	
serial number	Re number	diameter d (cm)	$C_w = \dfrac{W}{\dfrac{\gamma w^2}{2g} \dfrac{\pi d^2}{4}}$	with dimension	dimensionless form
1.	<2	<0.015	24/Re	$w = \left(\dfrac{\gamma - \gamma_l}{\gamma} \right) \dfrac{gd^2}{18v}$	Re/Fr = 18
2.	2÷744	0.015÷0.21	18.2/Re$^{0.682}$	$w = \sqrt[4]{\dfrac{g^3 d^5}{2209 v^2}}$	Re/Fr$^{3/2}$ = 47
3.	744÷1380	0.21÷0.72	0.366 We/Fr	$w = \sqrt{\dfrac{3.64 g\sigma}{\gamma d}}$	We = 3.64
4.	>1380	>0.72	2.61	$w = \sqrt{0.51 gd}$	Fr = 0.51

$v = \text{kinematic viscosity} = \dfrac{\eta}{\varrho} = \dfrac{\eta \cdot g}{\gamma}$

$d = \text{dimeter of sphere of identical volume}$

18*

Table 12.3. Rising velocity of bubbles in water

Ranges	1.	2.	3.	4.
Limits of validity, d (cm)	0.0—0.015	0.015—0.21	0.21—0.72	>0.72
Water temperature (°C)	rising velocity, w(cm/s)			
0	$3080\ d^2$	$193\ d^{5}/_{4}$	$16.6/\sqrt{d}$	$22.6\ \sqrt{d}$
10	$4260\ d^2$	$226\ d^{5}/_{4}$	$16.4/\sqrt{d}$	$22.6\ \sqrt{d}$
20	$5400\ d^2$	$256\ d^{5}/_{4}$	$16.2/\sqrt{d}$	$22.6\ \sqrt{d}$

12.7.2 Variables OC and $K_L a$

As is known, the oxygen input rate of air-blower systems is characterized by the oxygen-uptake capacity OC (g/m³h) and by the extended mass transfer coefficient K_L (h⁻¹):

$$OC = K_L a C_s = \frac{K_L A}{V} C_s \qquad (12.11)$$

[notations are those of eq (12.1)].

Several researchers found that OC and $K_L a$ vary as functions of the turbulence and that these variables are, to a certain extent, applicable for characterizing the magnitude of turbulence in an indirect way. This is implied in the relationship by Danckwertz, who defined K_L as

$$K_L = \sqrt{Dr}, \qquad (12.12)$$

where D is the molecular diffusion constant and r is the renewal rate of the diffusive surface.

Table 12.4. Relationship between pore size and bubble diameter (after Seeliger 1949)

Pore size D (mm)	Horizontal bubble diameter d (mm)	Ratio d/D (mean value)
0.200—0.380	2.2	8
0.130—0.160	1.2	9
0.045—0.070	0.8	14
0.025—0.035	0.5	17
0.020—0.030	0.5	20

In addition to D, the turbulent diffusion constant D_T can also be taken into account on the basis of the theory of Kishinevski (in Horváth 1984):

$$K_L = \frac{2}{\sqrt{\pi}} \sqrt{(D+D_T)r} .$$ (12.13)

In aeration basins $D \ll D_T$ in general cases.

Eventually, r is a function of turbulence. Similar conclusions can be drawn on the basis of the research results of Dobbins (1964) and others. Considering further that $a = A/V$ is the specific surface area, which is also significantly affected by turbulent mixing, then it can be concluded that both $K_L a$ and OC are in close relationship with the intensity of turbulence.

12.7.3 Coefficient α_{OC}

Suschka (1971) in carrying out experiments in basins with blown-air aeration, proved that the coefficient $\alpha_{oc} = OC_{sewage}/OC_{tap\ water}$ depends, in addition to the type of sewage, on the method and intensity of aeration. He found that α_{oc} is an exponential function of the blown-air discharge. The exponents of air discharge are negative numbers of an order of magnitude of one-tenth or one-hundredth (for waste waters containing phenols, the exponent was of positive sign).

In carrying-out experiments in the Hungarian Water Resources Research Centre (VITUKI), we found that the above statements can be considered of general validity, i.e., they are also valid for aeration systems other than the blown-air type (Horváth 1966a). In general, it may be stated that parameters affecting the intensity of aeration will affect the value of α_{oc}. For example, the immersion depth and the speed of rotation of aeration rotors influence the value of α_{oc}.

12.7.4 Dimensionless relationships

The relationships between mass transfer and hydraulic characteristics are frequently presented in (theoretical or empirical) dimensionless form. Some examples for various aeration basins will be presented below.

Based on the work of other authors, Eckenfelder (1970) characterized blown-air aeration systems by relationships between the Sherwood, the Reynolds, and the Schmidt numbers, as shown in *Fig. 12.19*:

$$\text{Sh } h_r^{1/3} = \frac{K_L d_B}{D} h_r^{1/3} = \left(\frac{d_B v_B}{v}\right)\left(\frac{v}{D}\right)^{1/2} = \text{Re Sc}^{1/2},$$ (12.14)

where h_r is the depth of air blowing, and d_B is the average bubble size. The range of validity of this expression is $50 < \text{Re} < 500$; $0.18 < h_r < 3.65$ m.

The present author carried out extensive experiments with aeration systems with the aim of improving their efficiency. The scale-model studies made for the INKA-

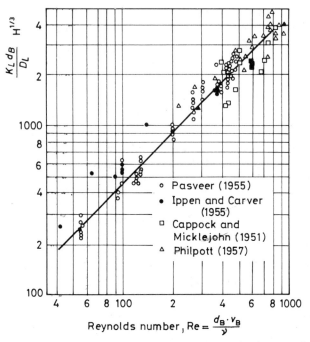

Fig. 12.19. Dimensionless correlation for blown-air aeration basins, in the case of clean water (after Eckenfelder 1970)

type basin of the STP of Hatvan, Hungary, yielded the following relationship (Horváth 1966a, 1970):

$$\frac{K_L a d_f}{v_B} = 1.048 \times 10^{-2} \left(\frac{h_r}{d_f}\right)^{0.7} \left(\frac{G d_f}{v_B}\right)^{1.12}, \tag{12.15}$$

where d_f is the diameter of perforations on the blown-air strainer (m); G is the input air discharge per unit effective volume of the basin (m³/m³h); and the range of validity is

$$4.42 \times 10^{-6} < K_L a d_f / v_B < 1.76 \times 10^{-5}$$

$$125 < h_r / d_f < 220.$$

A further example is shown in *Fig. 12.20*, in the case of a basin aerated with a special rotor of the horizontal axis type (Horváth 1966a). The dimensionless relationship yielded the critical Reynolds number that separates laminar and turbulent conditions. In this case $Re_{crit} = 3 \times 10^5$. Thus, the relationship between the variable $K_L a$, characterizing the rate of oxygen uptake, and variables relevant to the operational conditions is of different form for the two different flow conditions. In the turbulent range, the rate of oxygen uptake is not affected by the Re number (e.g., the curves take the form of horizontal straight lines).

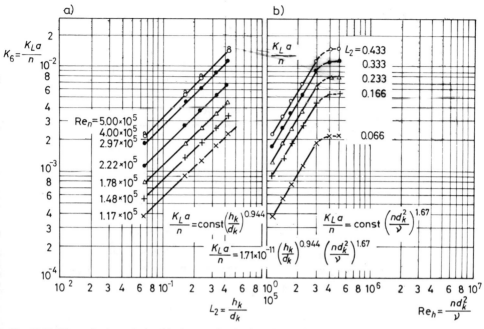

Fig. 12.20. Dimensionless relationship for aeration basins with horizontal-axis rotors, in the case of clean water

Finally, it must be emphasized that the above-presented relationships corres-pond, exclusively, to the given experimental conditions, or to those of similar systems.

12.8 Cell activity and flow

12.8.1 Oxygen concentration on the surface of activated sludge flocs

Pasveer (1956) carried out extensive experiments to determine the effect of turbu-lence on oxygen input and uptake rates. As shown in *Fig. 12.21/a,* he found that the active floc volume and its O_2 concentration decrease drastically with the decrease of floc size. It was also found that the relationships shown in the Figure are not affected by the extent of turbulence. Nevertheless, the O_2 concentration measur-able at the surface of the flocs depends (as shown in *Fig. 12.21/b*) on turbulence (Pasveer, however, did not characterize turbulence numerically). As indicated by the Figure, the O_2 concentration is around 90% at the surface of the flocs when tur-bulence is intensive, while it becomes independent of the floc size in the range of

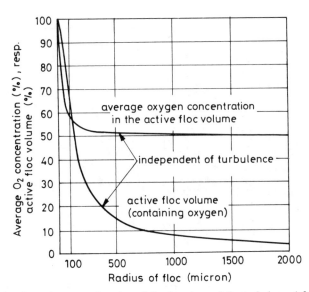

Fig. 12.21/a. Active floc volume as a function of the floc size and the turbulence (after Pasveer 1956)

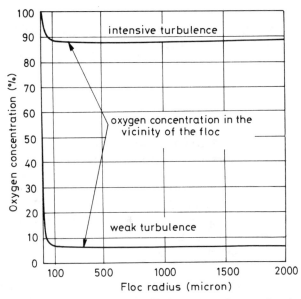

Fig. 12.21/b. Oxygen concentration on the surface of activated sludge flocs as a function of the floc size and the turbulence (after Pasveer 1956)

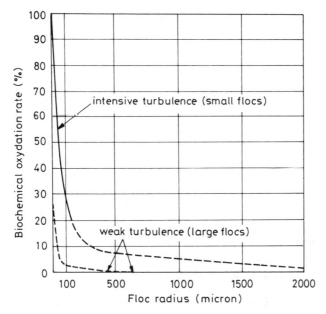

Fig. 12.21/c. Biochemical oxydation rate as a function of the floc size and the turbulence (after Pasveer 1956)

$r > 100\,\mu$. As contrasted to this, the active floc volume and the O_2 concentration within the flocs depend on the floc size within a wider range and become independent only when $r > 1000\,\mu$. It was emphasized that "micro turbulence" (velocity pulsation within a small volume) and "macro turbulence" (flow within a larger volume where the relative displacement in the vicinity of flocs is still small) are clearly distinguishable. The former effectively influences mass transfer conditions, while the latter represents convective flow and mass transfer.

12.8.2 Rate of biochemical oxydation

As a continuation of the above line of thought, Pasveer constructed *Fig. 12.21/c* that describes the rate of biochemical oxydation. It is indicated that the biochemical oxydation rate greatly depends on floc size, as well as on intensity of turbulence. The experiments carried out in relation with the above Figures yielded the well-known theorem of Pasveer (1956), the principle of high-rate activated sludge systems. The rate of biochemical oxydation is usually high, with the oxygen input rate OC being the limiting factor. Consequently, with increasing OC and turbulence, the loading rate of activated sludge plants can be increased. On the other hand, the volume of aeration basins — as compared to low-rate systems — can be significantly decreased. Comparing Pasveer's various graphs, it may be stated that the active floc volume, the O_2

concentration measurable at the floc surface and thus the rate of biochemical oxydation can — when assuming intensive turbulence — be best increased in the range of
$r < 1000\,\mu$.

12.8.3 Cell activity and reaction rates

Hartmann and Laubenberger (1968) investigated, under laboratory conditions, the
variation of the activity of the biological film and of the activated sludge flocs as
functions of the extent of turbulence, i.e., as a function of the Reynolds number. They
demonstrated the biological film grown on the inner surface of the wall of a glass tube
of 1.0 m length and 1.0 cm diameter. The experimental conditions were as follows:
mean flow velocity in the tube, $vk = 14$—107 cm/s; temperature, 22—23 °C; and the
kinematic viscosity of the fluid 1.3×10^{-6} m²/s. Assuming the validity of the Michaelis–Menten model of reaction kinetics, Hartmann (1967) determined the maximum reaction rate r_{max} and the Michaelis constant K_m for various flow velocities.
The results are shown in *Fig. 12.22*. The third model variable is the organic nitrogen
percentage, in dry weight, of the biological film.

The main conclusions were as follows: a) flow velocity — i.e., is the value of the
Re number — affects the reaction rates of the aerobic biological system. This effect
is mainly in the form of affecting the transfer of nutrients (substrate) towards the
micro organisms. The transport of nutrients might be the limiting factor in the chain
of reactions (the minimum principle of Liebing); b) turbulence affects oxygen transfer
as well; and c) the kinetic parameters are not constants but are affected by the flow
conditions. In this context, it should be noted that environmental conditions (temperature, pressure, flow velocity, turbulence, etc.) are not explicitly expressed in the

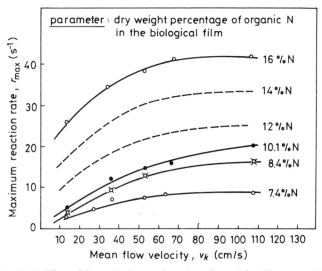

Fig. 12.22. Effect of flow velocity on the value of r_{max} (after Hartmann 1967)

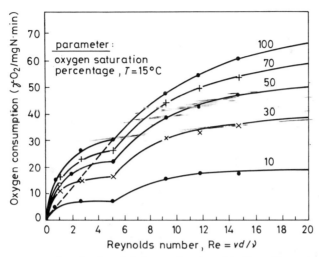

Fig. 12.23. Oxygen consumption of activated sludge as a function of the Reynolds number (after Hartmann and Laubenberger 1968)

reaction kinetic equations. Implicitly, however, these variables are reflected in the value of the kinetic coefficients.

The activity of activated sludge was analysed by the above authors in sludge suspensions flowing in a tube. The experimental conditions were as follows: $v_k = = 5{-}100$ cm/min and $T = 15, 20, 25, 30$ °C. Bioactivity was expressed in the form of oxygen consumption (γO_2/mg N/min, where N is the organic-nitrogen content of the floc). The results are shown in *Fig. 12.23*. The main conclusions are:

a) Turbulence has a double effect (hence the two well-distinguishable ranges of the curves in the Figure); partly it splits up the flocs, and partly it accelerates the transfer across the boundary layers. Consequently, oxygen consumption increases with the increase of Re. *b*) The disaggregation of flocs might be started already in the laminar range. Above Re ≈ 5000, the characteristics of flow do not significantly affect the physical properties of activated sludge flocs. *c*) Mass exchange across the boundary surface (because of the turbulence) between bacterial cells and sewage water takes place mainly in the domain Re $= 5000{-}9000$. *d*) Oxygen transport towards the surface of the cells might be an important limiting factor of biological decomposition rate.

Figure 12.24 illustrates that sludge aeration might also be affected by the extent of turbulence. By decreasing the speed of rotation of the magnetic mixer, the time variation of oxygen concentration will show an irregular pattern. Resetting the original speed of rotation, the oxygen-uptake curve regains its original regular shape. This effect might be due to the inaccurate measurements by the oxygen electrode, in the low ranges of turbulence.

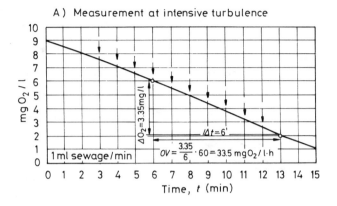

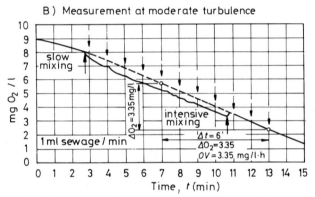

Fig. 12.24. Effect of turbulence on the measurement of sludge aeration (after Kayser 1967)

12.9 Turbulence and the intensity of mixing

Reviewing the relevant literature, one might conclude that the characterisation of turbulent flow in sewage-treatment facilities has been done mostly on a qualitative basis only (we refer to the work by Pasveer 1956). There are, however, some publications where attempts were made to make use of theoretically derived turbulence characteristics in the field of waste-water treatment technology. The work of Kalinske (1958) belongs in this category. The pioneering work of this author in the thirties gave rise to the practical application of the statistical theory of turbulence in the field of waste-water treatment. In his work Kalinske analysed the turbulence problems of aeration basins, pointing out that parameters of turbulence, such as the spectrum of turbulence or the turbulent dispersion coefficient, can be made use of also in the field of waste-water treatment technology. He gave a wide overview of the problem, referring to the works of Taylor, Prandtl and Kármán (in Horváth 1984).

In Hungary, the author of this book has carried out experiments to analyse the conditions of turbulence in aeration basins (Horváth 1984) and to determining the conditions of scale-modellability (Horváth 1970).

In addition to the intensity of turbulence, the extent of mixing is also a characteristic feature. These investigations were mainly based on throughflow characteristics (breakthrough waves). By analysing the distribution of retention times, the conditions of the deviation from ideal tube and tank reactors can be determined. In this context, the works of Ottengraf and Rietema (1969) and Milbury et al. (1965) can be mentioned, in which the simultaneous treatment of hydraulic and reaction kinetic problems were attempted.

12.10 Hydraulic design of aeration basins

12.10.1 Characteristic flow velocities

Bottom-flow velocities in aeration basins should be sufficiently high to avoid the settling and anaerobic decomposition of sludge. The minimum required bottom-flow velocity is 10—15 cm/s, while the most favourable velocity range is 25—35 cm/s. It is to be noted that in hydraulically favourable aeration basins and in the case of light sludges, even velocities lower than 10 cm/s will cause undesirable depositions. The mean throughflow velocity, as calculated from the hydraulic loading, is generally in the range of 0.5—2.5 cm/s, which is significantly lower than the above domain. Consequently, the role of the various aeration devices, and that of the flow velocities induced by their "pumping capacity", is extremely important from the hydraulic point of view. An appropriate geometric design might help to avoid the occurrence of dead flow zones. In this context, the main geometric dimensions of the structure, as well as the dimensions of the diversion baffles are also of importance.

In recirculation channels and pipes, the desirable flow velocity is 50—60 cm/s.

12.10.2 Hydraulic (volumetric) loading rates

The hydraulic load T_h on a certain activated-sludge unit is calculated as the sewage inflow discharge through a unit volume of the basin (m³ sewage/m³ · day of the basin volume):

$$T_h = \frac{Q_v}{V} = \frac{24}{t_s} = 24D \quad (\text{m}^3/\text{m}^3 \cdot \text{day}), \tag{12.16}$$

where V is the useful volume of the basin (m³); Q_v is the sewage inflow rate (m³/h); D is the dilution rate (h⁻¹); $t_s = V/Q_v = 1/D$ the calculated mean retention time (h) and 24 a conversion factor (h/day).

The hydraulic load is a variable characterizing the quantity of sewage arriving. For example, if $Q_v = 500$ m³/h and $V = 1000$ m³, then $T_h = 12$ m³/m³ · day.

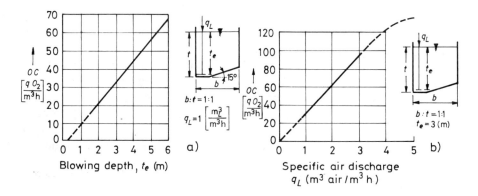

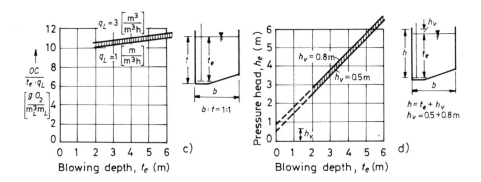

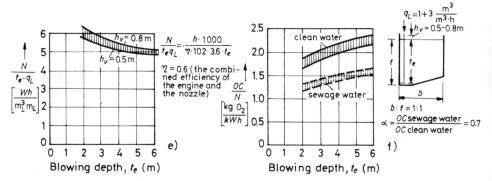

Fig. 12.25/a–f. Design guide for deeply aerated aeration basins (von der Emde 1969)

The *BOD* load (the organic load) can be calculated on the basis of the hydraulic load, by multiplying it with the *BOD* concentration in the inflow:

$$T_b = T_h C_0 = \frac{24 C_0}{t_s} \quad (\text{kg } BOD_5/\text{m}^3 \cdot \text{day}) \tag{12.17}$$

and the decomposed fraction of the *BOD* load is obtained as:

$$N_b = T_b \eta_{BOD} = T_b \frac{C_0 - C_e}{C_b} = T_h(C_0 - C_e) = 24 \frac{C_0 - C_e}{t_s}, \tag{12.18}$$

where C_0 and C_e are the influent and effluent *BOD* concentrations, respectively. For example, if $T_h = 12$ m³/m³ · day and $C_0 = 300$ mg/l = 0.3 kg/m³, then $T_h = 3.6$ kg/m³ · day; further, if $C_e = 30$ mg/l = 0.03 kg/m³, then $N_b = 3.24$ kg/m³ · day.

12.10.3 Some design aids

Figure 12.25/a—f presents design aids for deeply aerated aeration basins. Explanatory sketches facilitate the use of these graphs. In respect to *Fig. (d)*, it is to be noted that pressure head h is composed of aeration depth plus the head loss ($h = = t_e + h_v$). The head loss is further composed of the resistance of the air filter (if any), the head loss of the pipe lines and the entrance loss of the diffusor device (i.e., at the pores of the ceramic diffusor head). The characteristic relationships of fine-bubble aeration basins are shown in *Fig. 12.25/g*.

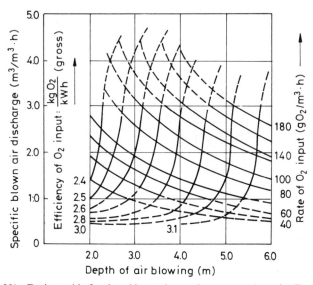

Fig. 12.25/g. Design guide for deep-blown-air aeration systems (von der Emde 1969)

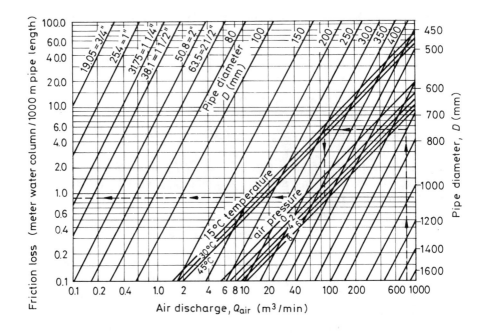

Air pressure in water column (m)

Conversion factors of head loss for various types of pipe			
type of pipe	Pipe diameter		
	3/4"- 21/2"	80-800	>800mm
Seamless steel pipe	0.80-0.95	0.84-0.93	0.82-0.92
Zinc-plated pipe	1.00		
Cast-iron pipe (uncoated)		1.18 -1.10	1.16 -1.10
Cast-iron pipe (with bitumen lining)		1.00	1.00
Concrete pipe (smooth)			1.20 -1.14
Concrete pipe (rough)			2.10 -1.85

Fig. 12.26. Nomogram for calculating the friction coefficient of pipe conveying air (after Knop, Bischofs-berger and Stalmann 1964)

Figure 12.26 presents a graph for determining the friction losses in air-supply pipes, thus allowing the design of the aeration system.

The nomogram of *Fig. 12.27/a* aids the calculation of air discharge needed for blown-air aeration. With the original notation of Kalbskopf (1966b): $L = O_v/O_E$ is the air quantity (N m^3/m^3) needed for aerating 1 m^3 sewage water, where $O_v = kB =$ = oxygen consumption (g/m^3); $O_E = a_s t =$ the oxygen input (gN/m^3); k is the factor of oxygen utilization (g O$_2$/m^3 · h); B is the BOD_5 value of the arriving sewage

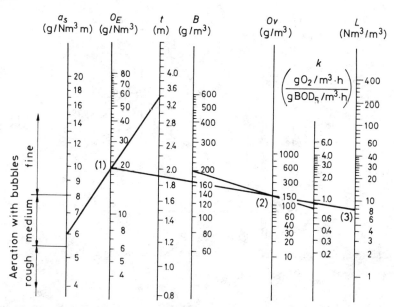

Fig. 12.27/a. Nomogram for calculating the air requirement of blown-air aeration basins (after Kalbskopf 1966)

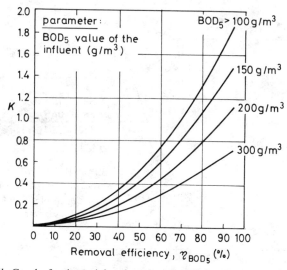

Fig. 12.27/b. Graphs for determining the value of coefficient k (after Kalbskopf 1966)

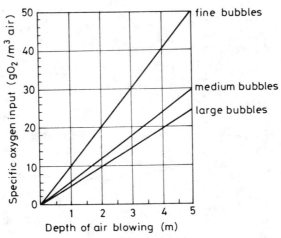

Fig. 12.27/c. Specific oxygen input rate as a function of the blowing depth (after Kalbskopf 1966)

(g/m^3); a_s is the specific oxygen input rate $(g/N\ m^3 \cdot m)$ and t is the depth of air blowing (m). The steps of using this nomogram are: a) the value of coefficient k is obtained from *Fig. 12.27/b* in function of B and the desirable removal efficiency; b) knowing a_s and t, O_E is obtained; c) O_B is determined on the basis of B and k; d) finally, knowing O_E and O_v, the value of L can be calculated. The value of a_s can be obtained from *Fig. 12.27/c.*

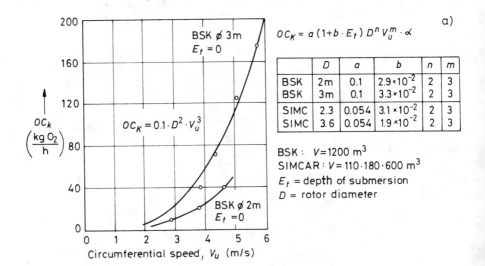

a)

$$OC_k = a\,(1+b \cdot E_t)\, D^n V_u^m \cdot \alpha$$

	D	a	b	n	m
BSK	2m	0.1	2.9×10^{-2}	2	3
BSK	3m	0.1	3.3×10^{-2}	2	3
SIMC	2.3	0.054	3.1×10^{-2}	2	3
SIMC	3.6	0.054	1.9×10^{-2}	2	3

BSK: $V = 1200\ m^3$
SIMCAR: $V = 110 \cdot 180 \cdot 600\ m^3$
E_t = depth of submersion
D = rotor diameter

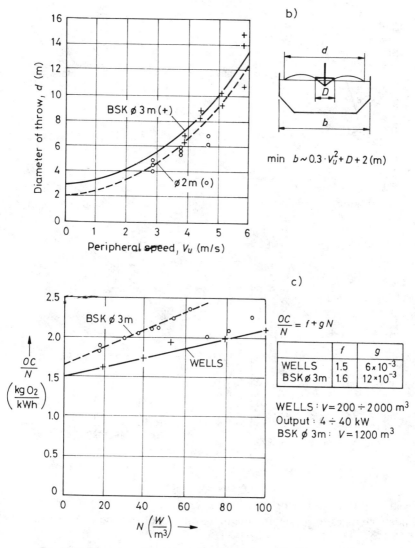

Fig. 12.28. Nomogram for designing vertical-axis aeration devices

In *Fig. 12.28*, relationships are shown for some well-known aeration facilities of the vertical-axis type (BSK, Simcar, Wells). Special attention is drawn to *Fig. (b)* that allows the calculation of the required basin width b as a function of the effective distance of rotor-driven flow (after von der Emde 1969).

A nomogram for designing the Hungarian-made aeration turbine VARIMIX is presented in *Fig. 12.29* (patent of the Vízgépészeti Vállalat in Hungary, FRG and

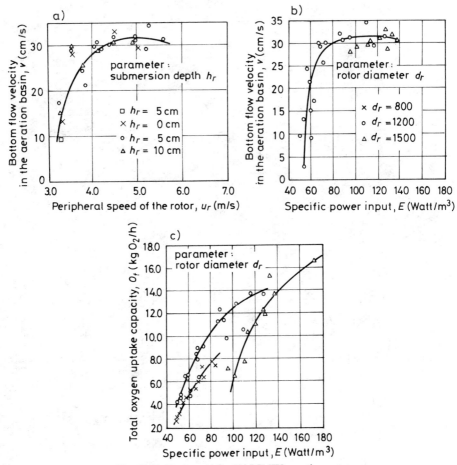

Fig. 12.29. Design aid for VARIMIX aeration rotor

Sweden; inventors I. Horváth, L. Nagy and I. Bánvölgyi). Data in *Table 12.5* are guideline values for designing the main dimensions of the structure and the aeration facilities.

12.10.4 Recirculation

Based on experiments in the fermentation industry, Herbert (1961) defined, using the balance equation, the dimensionless recirculation coefficient A in the activated-sludge system:

$$A = 1 + R_Q(1 - R_G), \qquad (12.19)$$

where

$$R_Q \triangleq \frac{Q_R}{Q_v} \quad \text{and} \quad R_G = \frac{G_R}{G_L},$$

Table 12.5. Main data and dimensions of VARIMIX rotors and aeration basins

Description	Notation	Unit	Value
Rotor diameter	d_r	[mm]	800—1000—1200—1500—2000
Submersion depth	h_r	[cm]	0—10; 5.0
Peripheral speed	u_r	[m/s]	4.0—5.5; 4.5—5.0
Water depth	h	[m]	2.5—3.0; max 3.5—4.0
Rotor diameter/basin width	d_r/d_m	—	0.1—0.2
Water depth/basin width	h/d_m	—	
if $d_m < 5$ m			0.6—0.7
if $d_m = 5 \div 10$ m			0.3—0.4
if $d_m > 10$ m			0.35
The maximum number of series-connected units in cascade-type arrangement		db	7

and where Q_v and Q_R are the influent sewage discharge and the recirculation discharge, respectively; G_L and G_R are the suspended organic-solid content of the aeration basin and the recirculated flow, respectively. It can be proved that the most favourable condition for activated-sludge plants refers to $A=0$. Consequently, the design equation of recirculation will be:

$$R_Q = \frac{1}{R_G - 1} = \frac{G_L}{G_R - G_L}. \tag{12.20}$$

The nomogram expressing the above equation is shown, as a design aid, in *Fig. 12.30* (Horváth 1984).

Thus, the ratio of R_Q (or $R\%$) to R_G directly expresses the role of recirculation. R_Q has, in addition to sludge transfer, also an important hydraulic role, affecting the distribution of retention times. The hydraulic role of the other directly affecting parameter, R_Q, is less significant (only affecting viscosity and density at higher sludge concentrations). However, R_Q has an important role in the reaction kinetics and the related processes.

12.10.5 Calculations related to special aeration devices

Gravitational aeration is relatively seldom used in waste-water treatment technology. This type of aeration might be provided by cascade aeration or by chutes. Vertical sprinkling or spraying, in one or more steps, can also be used for this purpose. The basic formulae for the hydraulic design of the latter are:

depth and time of free-fall are expressed as

$$h = \frac{1}{2} g t^2 \qquad t = \sqrt{2h/g}. \tag{12.21a—b}$$

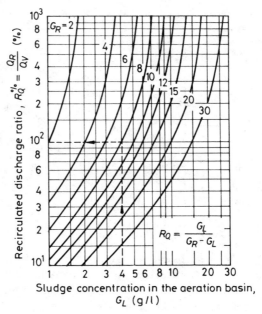

Fig. 12.30. Design of recirculation in activated sludge systems

If depth h is divided into n stages then

$$t = n\sqrt{2h/ng} = \sqrt{2nh/g}\ . \tag{12.22}$$

This solution can be recommended in cases when a sufficient pressure head is available and the area available for the structure is small. The usual height of praying is 1—3 meters. According to eq (12.22), time t can be increased to $\sqrt{n}$-times the original, by splitting the process into stages.

Nozzle aeration

The hydraulic principles of fixed aeration nozzles are based on the physical formulae of inclined throw (*Fig. 12.31*).

The height of rise of throwing at an angle α is:

$$h_r = \frac{1}{2} g t_r = C_v^2 h \sin^2 \alpha. \tag{12.23}$$

The time of rising (expressed as the total time of throw):

$$t_r = \frac{1}{2} t = v \sin \alpha/g = C_v \sqrt{2h/g} \sin \alpha, \tag{12.24}$$

and the pressure head required is:

$$h = g t_r^2/(2 C_v^2 \sin^2 \alpha). \tag{12.25}$$

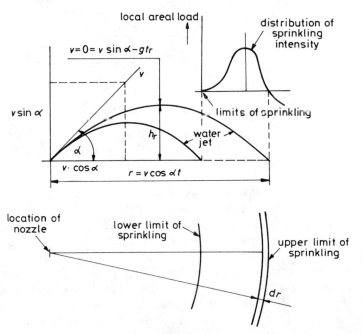

Fig. 12.31. Scheme for inclined water jets from fixed nozzle (the case of inclined throw) (after Fair, Geyer and Okun 1968)

The effective radius of spraying (neglecting the wind effect):

$$r = 2vt, \cos \alpha = 2C_v^2 h \sin 2\alpha. \tag{12.26}$$

The initial velocity of inclined throw is

$$v = C_v \sqrt{2gh} . \tag{12.27}$$

Water discharge through the nozzle:

$$Q = \mu F \sqrt{2gh} . \tag{12.28}$$

In the above equations, C_v is the velocity coefficient (≈ 0.90—0.95); μ is the discharge coefficient (for sharp-edged openings $\mu = 0.6$—0.7, for rounded openings $\mu = 0.8$—0.85, and for nozzles $\mu = 0.85$—0.95) and F is the total area of the openings. The effect of wind can be considered approximately by the following relationship:

$$l = 2C_D v_w t_r, \tag{12.29}$$

where l is the length of delivery; v_w is the wind velocity and C_D is the drag coefficient (≈ 0.6). (The hydraulic principles of moving sprayer devices are included in Chapter 13. The design methods of perforated effluent pipes can be found in hydraulic manuals.)

13 Trickling filters

Trickling filters are columns filled with filter medium. Filter materials might consist of crushed stone, slag, plastics, etc. The biological decomposition of contaminants is carried out, with the presence of appropriate quantities of nutrients and dissolved oxygen, by microorganisms present in the biological film formed on the surface of the filter medium.

Flow conditions within the filter affect the technological processes as a result of the following two essential factors: a) the flow velocity in the liquid-film surrounding the filter medium, i.e., the distribution of surficial velocities, and b) the retention time of the sewage water, i.e., the distribution of retention times. In addition to these factors, the following hydraulic conditions exert an influence on the operation of trickling filters: the flow of air through the column, the so-called "chimney-stack" effect; the relationship between surface loading rate and washing rate; the recirculation rate: the distribution of sewage on the surface of the filter medium, and the hydraulic properties of the revolving sprinkler.

Basically, it is the flow velocities that determine the transport processes (the input rate of the substrate and the dissolved oxygen, the removal rate of metabolic products, the flushing rate of the biological film, etc.).

The retention time, will, at the same time, significantly affect the rate of biological reactions in the filter. It follows from the above considerations that theoretical and experimental investigations of the hydraulic processes within the filters are mostly focussed on the elucidation of the role of the above factors.

13.1 Velocity and retention time

The following methods of investigation are used:

a) characterization of the environment of a single filter element (a particle) — point investigation;

b) characterization of the processes along a vertical series of filter elements — vertical investigation;

c) characterization of flow whithin a set of filter particle — spatial investigation.

In each of the above three cases the investigation might be focussed on the determination of the above-mentioned two basic variables — the flow velocity and the retention time. In ideal cases — mostly for mathematical treatment — series or sets of spherical elements are considered.

a) Single filter element — spherical particle

This analyisis is of mostly theoretical importance. The relationship between the most important variables of the vicinity of a particle can be determined in this way. The mathematical investigations by Franzini and Hassan (1964) with the hydrodynamic analysis of thin liquid films can be mentioned as examples. These authors have shown that the velocity distribution in a fluid film can be best approximated by a second order parabola — where $v=0$ at the surface of the particle and $v=v_{max}$ at the surface of the fluid.

According to Howland (1958) and associates, the average retention time in the liquid film around a spherical particle of radius R is:

$$t_a = 2.6(2\pi)^{2/3}\left(\frac{3\eta}{\gamma}\right)^{1/2}\frac{R^{5/3}}{Q^{2/3}},\tag{13.1}$$

where Q is the rate of flow within the film; γ is the specific weight of the flowing media and η is the kinematic viscosity of the liquid.

b) Characterizing the vicinity of a vertical series of particles

Bloodgood et al. (1959) described the flow conditions of trickling filters under the assumption of a vertical series of filter elements. The results of their investigations are of mainly theoretical importance and can be used for the determination of the average retention time. The results of Bloodgood et al. can be presented in the form of the following relationship:

$$t_a = \text{const}\,\frac{n}{Q^a},\tag{13.2}$$

where n is the number of spheres in the series; Q is the rate of flow within the fluid film; $\text{const}=0.418$ and $a=0.655$ for a particle diameter of 2.54 cm; $\text{const}=2.67$, and $a=0.601$ for a particle diameter of 8.9 cm. This analysis does not reveal the role of the lateral interaction of the elements.

c) Characterizing the flow whithin a set of particles

The most utilizable practical method of investigating the flow conditions of trickling filters relies on the consideration of a set of particles and their interactions.

Sinkoff et al. (1959) investigated the ideal case when the filter medium consists of spherical particles. Using theoretical dimension–analytical considerations and experimental results, this author found that the retention time is affected by the following variables:

$$t_a = f(H, S, Q, v, g),$$ (13.3)

where H is the depth of the filter medium; S is the specific area of the medium = the total surface of the filter elements per volume of filter medium; g is the acceleration of gravity and v is the kinematic viscosity of the flowing medium.

Interpreting the measurement data graphically — for a filter medium consisting of glass spheres ($\emptyset$ 0.8, 1.25, 2.5 cm diameters) — he obtained the following results:

$$t_a = 3.0 H \left(\frac{S}{Q} \right)^{0.83} \frac{v^{1/2}}{g^{1/3}}.$$ (13.4)

Conclusions drawn on the basis of relationships for the retention time are: there is a linear relationship between the average retention time and the depth of the filter medium. This is also indicated by eq (13.2), where the number of particles n replaces depth H. This relatively trivial relationship is also supported by eq (13.4). For a specific surface area S, that plays a role in this equation, it should be noted that there is only a limited possibility for increasing S, as this decreases the voids between the particles, which in turn results in clogging. The most important variable in practice is the discharge rate Q or the surface loading rate, that also affects the average retention (contact) time t_a. In addition to the above relationships, the following measurement results are worth mentioning: the exponent of Q in eq (13.2) is, according to different authors, as follows:

according to Bloodgood et al. (1959) $a = 0.65$
according to Burgess et al. (1961) $a = 0.408$
according to Gruhler (1962) $a = 0.69$
according to Howland (1958) $a = 0.666 = 2/3$
according to Meltzer (1965) $a = 0.75 = 3/4$
according to Sinkoff et al. (1959) $a = 0.53—0.82$

Comparing various literature data on the value of a, it is found that

$$0.408 < a < 0.83,$$

where the differences stem from those of the experimental conditions. In harmony with experimental results, the following formula can be recommended for practical purposes:

$$t_a = \text{const} \frac{H}{Q^{2/3}}.$$ (13.5)

The value of const can be determined experimentally.

13.2 Throughflow studies in trickling filters

In conjunction with the above considerations, some measuremental results on the experimental determination of the retention time and its distribution are presented below. As early as the beginning of this century, Dunbar, one of the pioneers of waste-water treatment technology, recognized the importance of retention time, when studying trickling filters. Based on experiments with an approx. 1.0 m high column, he found that significant removal efficiencies can be expected even in the case of short retention times. Blunk (1933b) also analyzed the changes of retention time in an experimental trickling-filter facility, with a depth variable up to 4.0 m. The retention time of sewage water varied, as a function of the particle size of the filter medium and the loading rate, between 1.0 and 5.0 hours.

Among the throughflow studies in trickling filters, the experiments by Pönninger (1965) are of outstanding importance. Some of his results will be presented below. *Figure 13.1* shows breakthrough curves determined by dosing uranin. The time variation of tracer concentration was determined as a function of the hydraulic loading rate (m³ per m³/day). Average retention times $t_1 \ldots t_6$ were determined as those corresponding to the vertical projection of the centre of "gravity" of the area under the breakthrough wave. The average retention time decreases with an increasing hydraulic load, along with a decrease of the treatment efficiency. The experimental conditions of the six experimental alternatives shown in the Figure are presented in *Table 13.1*, together with the data of three other experimental runs.

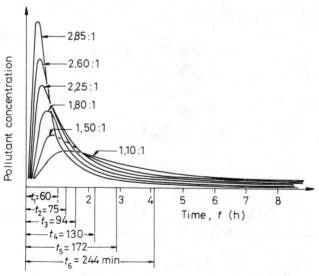

Fig. 13.1. Breakthrough curves of trickling filters at various hydraulic loading rates (after Pönninger 1965)

Table 13.1. Technological data of trickling filter experiments (after Pönninger 1965)

	Volumetric load (m³/m³·day)	BOD_5 conc. of the influent, C_0 (mg/l)	BOD load, T_b (g/m³·day)	Average retention time, t_s (min)	Surface loading rate, T_f (m/h)	BOD_5 conc. of the effluent, C_e	Treatment efficiency, $BOD\%$	Filter diemensions, $\varnothing$ and h (m)
	1	2	3	4	5	6	7	8
1	1.10:1	450	495	244	0.17	10	98	
2	1.50:1	450	675	172	0.23	10	98	$\varnothing$ 14.5
3	1.80:1	450	810	130	0.28	12	97.5	$h=3.7$
4	2.25:1	450	1010	94	0.35	16	96.5	
5	2.60:1	450	1170	75	0.40	21.5	95.5	
6	2.85:1	450	1283	60	0.44	25	94.5	
7	2.0 :1	450	900	58	0.125	34.4	92.5	$\varnothing$ 14.5 $h=1.30$
8	2.0 :11	450	900	55	0.31	28.8	93.7 (small particles)	$\varnothing$ 1.5 $h=3.70$
9	2.0 :1	450	900	44	0.31	43.0	90.5 (large particles)	$\varnothing$ 1.5 $h=3.70$

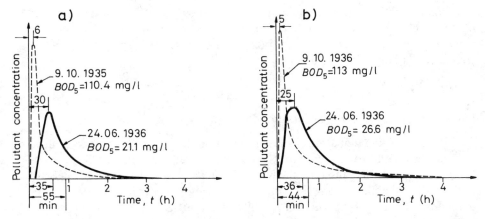

Fig. 13.2. Breakthrough curves of trickling filters with filter media of different particle sizes (after Pönninger 1965)

In *Fig. 13.2* breakthrough curves for small (7 cm) and large (20 cm) filter-medium particles are presented, for two points in time each. It was indicated that after two months operation, the breakthrough curves were shifted substantially to the left. Shorter retention times were mainly due to the growth of the biological film. Shorter contact times allowed of lower effluent BOD_5 concentrations as well. Comparing the effects of the size of the filter-medium particles, it was found that the smaller-size filter medium required 25% longer retention times for the same treatment efficiencies (7 cm diameter $-t_1 = 55$ min; 20 cm diameter $-t_2 = 44$ min; $t_1/t_2 = 1.25$).

Finally, in *Fig. 13.3* the experimental results of Schultze (1960) and Pönninger (1965) are compared, showing breakthrough curves for trickling towers *(1)* and

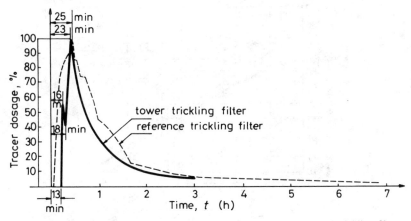

Fig. 13.3. Comparison of breakthrough curves of tower and traditional trickling filters

traditional trickling filters *(2)*. The breakthrough curves of the two alternatives are of similar character. To facilitate comparison, the tracer concentration is shown as percentage of the dosage concentration on the vertical axis of the Figure. Some characteristic points in time are:

		1	2
time of appearance of tracer	t_k (min)	13	3
time of maximum tracer concentration	t_m (min)	23	25.

It is worth noting that after 15 min of operation there was a smaller concentration peak on the breakthrough curve for the trickling tower. According to Schultz (1956), this phenomenon was caused by the water flowing downward along the inner surface of the wall of the filter (along this smooth surface both water and tracer move faster than along the irregular surface of the filter medium).

13.3 Loading rate of trickling filters

Surface loading rate

The surface loading rate of trickling filters can be given in the way generally used in sewage-treatment technology as

$$T_f = \frac{Q_v}{A_f} = T_h H_f \left(\frac{m^3}{m^2 h} \right), \qquad (13.6)$$

where H_f is the effective height of the filter column (m); A_f the horizontal cross-sectional area of the filter column (m^2); and T_h is the hydraulic loading rate of the filter (h^{-1}).

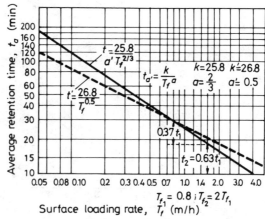

Fig. 13.4. Relationship between surface loading rate and retention time (after Pallasch and Triebel 1967)

In interpreting T_f, the time distribution of the sewage discharge rate can be taken into account by replacing Q_v by a fraction thereof, e.g., $Q_v/18$, $Q_v/16$, $Q_v/18$, etc., as a function of the size of the plant and the uniformity of the influent flow. The relationship between surface loading rate and the contact time is shown in *Fig. 13.4.* The relationship is, eventually, in harmony with eq (13.5).

If the recirculation water is also considered, then surface loading rate can be expressed as

$$T_{fR} = \frac{Q_v + R_Q Q_v}{A_f} = \frac{(1+R_Q)Q_v}{A_f}, \tag{13.7}$$

where R_Q is the recirculation rate.

For a given task, the values of T_f and T_{fR} can differ from each other significantly, and this difference must not be neglected. For low-rate systems, the surface loading rate is 0.05—0.25 m/h, while for high-rate systems the respective range is 0.6—1.5 m/h with or without application of recirculation. If, for example, $A_f = = 100$ m^2 and $Q_v = 2400$ m^3/day, then $T_f = 24$ m/day $= 1.0$ m/h.

Volumetric, hydraulic load

The presettled influent flow per unit of trickling filter volume is

$$T_h = \frac{Q_v}{A_f H_f} = \frac{T_f}{H_f} \quad \text{(m}^3/\text{m}^3 \cdot \text{day)}, \tag{13.8}$$

where $A_f H_f = V_f$ is the volume of the trickling filter.

If, for example, $T_f = 1$ m/h $= 24$ m/d and $H_f = 4$ m, then $T_h = 6$ m^3/m$^3 \cdot$ day.

The surface loading rate defined by the above relationship is an average value, since the distribution of water on surface A_f — assuming the use of a revolving sprinkler — is periodical. The instantaneous load on the various sectors of the filter changes with the position of the distribution device. It follows from this arrangement that the actual washing effect is more intensive than could be assumed on the basis of the value of T_f. The washing or flushing effect is significant from many points of view. This effect is related to: the continuous washing of parts of the biological film; to the transport of nutrients, dissolved oxygen and metabolic products; and to the continuous renewing of the active contact surface between the biological film and the sewage water, which could be favourable for absorption and adsorption processes, etc. Increasing the value of Q and T_f above certain limits may lead to unfavourable hydraulic effects. On the one hand, the average retention time t_a decreases intensively, while on the other, an increase of Q results in a changed character of throughflow such that the biological film is removed from parts of the surface of the filter medium, thus allowing the "free-falling" flow of sewage across the filter.

13.4 The role and extent of recirculation

Hydraulic effects of recirculation

The efficiency of treatment plants equipped with trickling filters can be increased by recycling some of the effluent water. The increasing efficiency is due to the combined effect of physical (hydraulic), chemical and biological processes.

From the hydraulic point of view, the effects of recirculation are usually favourable to the operation of trickling filters, and some of these are worth mentioning:

a) shock-like load variations are smoothed by recycling;

b) the spatial distribution of load becomes more uniform, thus facilitating the growth of the biological film. Some boundary surfaces that could have contributed less efficiently to the mass transfer and diffusion processes can be wetted and this involved in the operation;

c) the average loading rate of the filter can be increased due to the above-mentioned effects;

d) flushing rates can be increased — an important aspect in maintaining an equilibrium thickness of the biological film and for the appropriate day and night operation of water-distribution facilities (e.g., the revolving sprinkler);

e) increased flow rate increases turbulence which enhances mass transfer (oxygen input) and absorption processes.

The equalizing role of recirculation will be emphasised below, using an example from Pönninger (1965).

Consider a trickling filter receiving Q (m³/d) of presettled sewage water. The hourly peak load — without recirculation could be $Q/14$, while the hourly average flow is $Q/24$. *Figure 13.5* shows the diurnal variation of the sewage flow. In this Figure, the recirculated dilution water equals $Q/24$, the hourly average flow. Thus with recirculation the peak flow becomes:

$$\frac{Q}{14} + \frac{Q}{24} = \frac{38}{336}Q = \frac{19}{336}2Q, \tag{13.9}$$

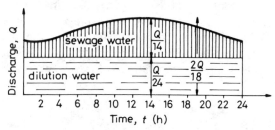

Fig. 13.5. The hydraulic effect of dilution by recirculation (after Pönninger 1965)

where $2Q$ is the total (influent + recycled) daily load on the trickling filter ($19/336 \approx$ $\approx 1/18 < 1/14$). The hydraulic efficiency is increased by about 28%.

The extent of recirculation

Treated effluent of concentration C_e is recycled during the treatment process, in order to achieve the appropriate treatment efficiency. (It is to be noted that in some cases deposited sludge is also being recirculated, similarly to the case of the activated sludge treatment.) Recirculation rate R_Q can be determined from the following balance equation when influent concentration C_e is diluted to a value C_{oR} with the total flow being $1 + R_Q/Q_v$:

$$Q_v C_0 - R_Q Q_v C_e = (Q_v + R_Q Q_v) C_{oR},$$

whence

$$R_Q = \frac{C_0 - C_{oR}}{C_{oR} - C_e}. \tag{13.10}$$

From the viewpoint of practical design, the calculation of resultant concentration C_{oR} arriving at the trickling filter can also be important. Rearranging eq (13.10), it is obtained that

$$C_{oR} = \frac{C_0 + R_Q C_e}{1 + R_Q}. \tag{13.11}$$

Example: $C_0 = 300$ mg/l; $C_e = 30$ mg/l and $R_Q = 1$, then $C_{oR} = 165$ mg/l.

13.5 Air flow through the trickling filter

There is a naturally induced air flow through the filter, due to the so-called "chimney-stack" effect. The air-flow velocity is determined by the difference between the ambient air temperature and that of the sewage water. Due to a pressure difference ΔP at the bottom level of the trickling filter, a static air flow is induced:

$$\Delta p = p_1 - p_2 = hg(\varrho_1 - \varrho_2),$$

where ϱ_1 and ϱ_2 are the air densities inside and outside the filter, respectively; and h is the height of the trickling filter.

During the cold season, the inner warm air flows upward. This means a counter flow as compared to the direction of sewage flow. In the summer months, the direction of air flow is reversed and matches that of the sewage flow.

In *Fig. 13.6* the relationship between air-flow velocity, the ambient air temperature and the sewage-water temperature is shown. Experience indicates that the linear relationship provides a good approximation. It is worth noting that a transient condition (e.g., air = 0) appears at a temperature difference of 2 °C, and not at a zero temperature difference, which could be expectable at first sight. This can be explained

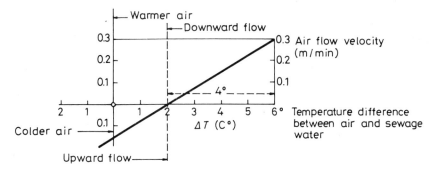

Fig. 13.6. Air-flow velocity as a function of the temperature difference (after Halvorson 1936)

by the higher moisture content of air flowing inside the filter. It is also found that at a temperature difference of 4 °C (as calculated from the intersection of the straight line and the horizontal axis) the air-flow velocity will be around 30 cm/s = 18 m/h. Considering an average surface loading rate of $T_f = 0.8$ m/h, the ratio of air-flow velocity to sewage-flow velocity is found as

$$v_{air} : T_f = 18 : 0.8 = 22.5.$$

Pöpel (1943) also found a linear relationship between temperature difference and the air-flow velocity:

$$v_{air} = 4.45(T_s - T_{air} - 1.88) \quad (m^3/m^2h = m/h). \tag{13.12}$$

According to the above expression, $v_{air} = 0$ corresponds to a temperature difference $(T_f - T_{air}) = 1.88$ °C, that relatively closely matches the value of 2 °C in Fig. 13.6. Finally, it should be noted that the air-flow conditions within a trickling filter are also affected by the external air pressure (e.g., by the wind) and by uneven warming-up (e.g., due to sunshine) which can induce secondary air currents within the filter.

13.6 Some hydraulic aspects of the operation of revolving sprinklers

The distribution of sewage water can be carried out by a revolving sprinkler device which can operate on the basis of the Segner-wheel principle, but can also be driven electrically.

Revolving sprinklers generally have 4 or 6 sprinkler arms, equipped with perforations or nozzles. The distance between perforations decreases with the distance towards the outside perimeter of the filter, in order to maintain a uniform sewage-water distribution. The momentum force of the water flowing out from the openings provides the driving force for the device. In order to ensure this operation, the openings should be located at the appropriate side of the sprinkler arm (i.e., on the side opposite to the direction of rotation).

The sewage flow to be distributed is:

$$Q = \mu n \frac{d^2 \pi}{4} \sqrt{2gH} , \tag{13.13}$$

where n is the number of perforations; μ is the discharge coefficient of the opening (the approximate value is 0.6—0.7); d is the diameter of the perforations and H is the pressure head over the axis of the perforated arms.

The pressure head needed for rotating the device, i.e., the head loss over the device, is proportional to the square of the discharge. On the other hand, the momentum force also changes with the square of the discharge, and thus the speed of rotation of the sprinkler depends on discharge Q. It is to be noted that the centrifugal force acts on the water flowing in the pipe-arms, and thus the centrifugal pressure head is obtained as

$$H_c = \frac{v^2}{2g} = \frac{\omega^2 r^2}{2g} , \tag{13.14}$$

where ω is the angular velocity. The distribution of the outflow from the openings, as determined by peripheral speed v (i.e., by H_c) varies with the radial distance r, a fact that should be taken into account in designing the distribution of the perforations. Eventually, at larger distances r, H_c increases, thus increasing the outflowing specific discharge (per unit distance of the pipe arm). *Figure 13.7* is presented for the illustration of eq (13.14).

Figure 13.8 shows the characteristic curves of a rotating sprinkler equipped with six pipe arms. The arms are grouped in pairs, thereby allowing the operation of fewer arms (one pair for example) at lower influent discharge rates. Finally, *Fig. 13.9* (after Rumpf n.d.) illustrates the common work-point of the revolving sprinkler and the pump. In this Figure, curves *1, 2,* and *3* are those for the constant geodetical level, the

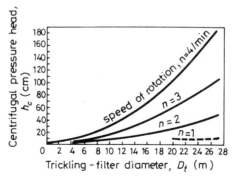

Fig. 13.7. Relationship between centrifugal pressure head, filter diameter and the speed of rotation (after Schwalbenbach 1964)

20*

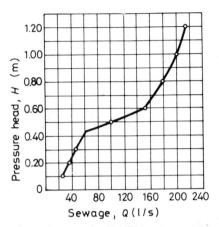

Fig. 13.8. Hydraulic characteristics of revolving sprinklers (after Rumpf)

head loss due to pipe friction and the total pressure head needed for operating the sprinkler device, respectively. The intersection of curve $Q(H)$ with curve *3* gives sewage flow rate Q_1 and the corresponding lifting head H_1. Curve *4* corresponds to a certain degree of choking (for example, to partial closure with a gate valve), when $Q_2 < Q_1$ and $H_2 < H_1$.

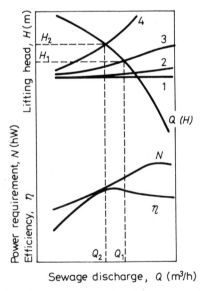

Fig. 13.9. Determination of the common work-point of rotating sprinkler and pump

13.7 Hydraulic design of water-distributor pipe arms

Ordon (1966) has dealt with the details of designing standing and revolving perforated sewage-water distribution devices. An example of the design of water-distributing revolving sprinklers for trickling filters will be presented below, after this author. The task is to determine the longitudinal changes of pressure, velocity and discharge, together with the respective energy and pressure profiles.

Consider a perforated pipe of length L and constant cross-sectional area, where the outflow through the perforations is q and the flow in the pipe is Q. The mean flow velocity and discharge Q change with the distance along the pipe, defining a steadily changing flow. Denote the discharge entering the pipe by Q_{in} and the discharge leaving it by Q_{out}. The scheme of operation is shown in *Fig. 13.10.*

Assuming that the decreasing rate of flow in the pipe is proportional to the second power of the distance, discharge Q_1, measureable at an arbitrarily chosen distance l_a, can be calculated on the basis of continuity considerations (from the balance equation) as:

$$Q_1 = Q_{in} - \frac{l^2}{L^2}(Q_{in} - Q_{out}). \tag{13.15}$$

The dynamic equation of fluid motion can also be given, utilizing the Chezy relationship, as:

$$I = \frac{dh_v}{dt} = \frac{Q_1^2}{A^2 C^2 R} = KQ_1^2, \tag{13.16}$$

where A is the cross-sectional area of the distributor pipe (m^2); R is the hydraulic radius of the pipe (m); and C is the velocity coefficient of Chezy (m$^{1/2}$ s^{-1}).

Head loss h_v can be obtained by combining the above two equations:

$$dh_v = K\left[Q_{in} - \frac{l^2}{L^2}(Q_{in} - Q_{out})\right]^2 dl \tag{13.17}$$

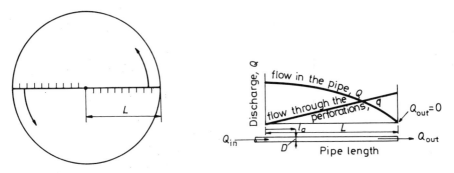

Fig. 13.10. Scheme of operation of revolving sprinklers (after Ordon 1966)

and by further rearrangement;

$$dh_v = KQ_{in}^2\left[1 - \frac{2l^2}{L^2} + \frac{2l^2 Q_{out}}{L^2 Q_{in}} + \frac{l^4}{L^4} - \frac{2l^4 Q_{out}}{L^4 Q_{in}} + \frac{l^4 Q_{out}^2}{L^4 Q_{in}^2}\right]dl.$$

Integrating between 0 and L the total head loss H_m is obtained as

$$H_m = KQ_{in}^2\left[L - \frac{2L}{3} + \frac{2LQ_{out}}{3Q_{in}} + \frac{L}{5} - \frac{2LQ_{out}}{5Q_{in}} + \frac{LQ_{out}^2}{5Q_{in}^2}\right] =$$

$$= \frac{8}{15}KLQ_{in}^2\left[1 + \frac{Q_{out}}{2Q_{in}} + \frac{3Q_{out}^2}{8Q_{in}^2}\right]. \tag{13.18}$$

For the water-distributor arms of revolving sprinklers $Q_{out}=0$, and for this condition

$$H_m = \frac{8}{15}KLQ_{in}^2 \approx \frac{8}{15}LI_{Q_{in}}, \tag{13.19a}$$

the approximate equivalence is based on eq (13.16), and $I_{Q_{in}}$ denotes the hydraulic gradient corresponding to Q_{in}.

According to Ordon (1966), for practical purposes, the following (rounded-off) formula, based on eq (13.19), can be used

$$H_m \approx 0.6 LI_{Q_{in}}. \tag{13.19b}$$

At a selected distance l_a, the head loss, as shown in *Fig. 13.11*, is

$$H_a = 0.6 LI_{Q_{in}} - 0.6(L - l_a)I_{Q_{in}}. \tag{13.20}$$

The above relationships do not yet account for the centrifugal force. Substituting $\omega = 2\pi n = 2\pi/T$ into eq (13.14), the centrifugal pressure head H_c is obtained as

$$H_c = \frac{(2\pi n)^2 l^2}{2g} = 2.02n^2l^2 = 2.02\frac{l^2}{T^2}, \tag{13.21}$$

where n is the speed of rotation of the distributor device (r.p.m.); and T is the time for one rotation (min).

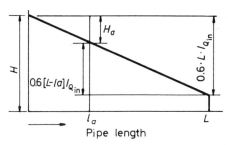

Fig. 13.11. Approximate changes of pressure along the pipe (after Ordon 1966)

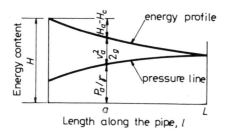

Fig. 13.12. Energy and pressure-head profiles along the water distributor pipe (after Ordon 1966)

A further task is the determination of the mean flow velocity and the respective head distribution in the pipe. Considering that $Q_a = A v_a$ eq (13.15) yields:

$$v_a = v_{in} - \frac{l_a^2}{L^2}(v_{in} - v_{out}).$$ (13.22)

Since $v_{out} = 0$, then

$$v_a = v_{in} - \frac{l_a^2}{L^2} v_{in} = v_{in}\left(1 - \frac{l_a^2}{L^2}\right) = \frac{Q_{in}}{A}\left(1 - \frac{l_a^2}{L^2}\right).$$ (13.23)

Thus, if $l_a = 0$ then $v_a = v_{in}$ and if $l_a = L$ then $v_a = 0$.

The above relationships help in constructing the head and energy profiles of the water-distribution device.

A schematic illustration of these hydraulically characteristic lines is presented in *Fig. 13.12*. At the beginning of the distributor pipe, the energy content is H which is decreased by subtracting head loss H_a and considering centrifugal head H_c. The latter increases the energy content and thus the difference $(H_a - H_c)$ should be subtracted from the original energy content H to obtain the energy line. The pressure profile of the distributor pipe is obtained by subtracting the velocity head $v_a^2/2g$ from the energy line. *Figure 13.12* also illustrates the variation of velocity head with the distance, as the difference between the energy and pressure profiles. As follows from the principle of the conservation of energy, the piezometric head at a given point a is obtained as

$$H = \frac{p_a}{\gamma} + \frac{v_a^2}{2g} + (H_a - H_c) = \text{const},$$ (13.24)

whence, after rearrangement and substitutions:

$$\frac{p_a}{\gamma} = H - H_a + H_c - \frac{v_a^2}{2g} = H - 0.6 L I_{Q_{in}} + 0.6(L - l_a)I_{Q_{in}} + 2.02\frac{l_a^2}{T^2} - \frac{v_a^2}{2g}.$$ (13.25)

Knowledge of pressure head $p_a/\gamma = H'$ allows the determination of the discharge flowing out through a perforation of cross-sectional area f:

$$q = \mu f \sqrt{2gH'}.$$ (13.26)

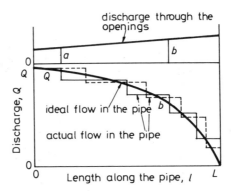

Fig. 13.13. Longitudinal profiles of flow in the pipe and flow through the perforations (after Ordon 1966)

It is expedient to compare eqs (13.13) and (13.26). In the former, μ is the average discharge coefficient and H is the average pressure head. In the latter, however, H' is the pressure head measured at the opening investigated.

Finally, *Fig. 13.13* illustrates the longitudinal variation of discharge over one arm of the water-distribution device. In the ideal case, initial discharge Q decreases to zero following a curve as shown in the Figure. This curve can be approximated with discrete steps with a desirable accuracy. Then the steps denote the location of perforations. The dashed step-line shown in the Figure marks an attempt to provide for the most uniform load onto the trickling filter. The solid line in the upper part of the Figure illustrates the longitudinal profile of the discharge flowing out through the perforations.

References

ALBRECHT, A. E. (1967): Aerated grit operation design and chamber. *Water and Sewage Works*, 9.
ANDERSON, N. E. (1945): Design of settling tanks for activated sludge. *Sew. Works J.*, 17, 50.
ARGAMAN, Y.—KAUFMANN, W. J. (1970): Turbulence and flocculation. *Proc. ASCE*, SA2 223—241.
BENDEL, H. L. (1966): Grit traps in sewage treatment. *Effluent and Water Treatment Journal*, 9.
BLOODGOOD, D. E.—TELETZKE, G. H.—POHLAND, F. G. (1959): Fundamental hydraulic principles of trickling filters. *Sew. and Ind. Wastes*, **31**, 3, 243—253.
BLUNK, H. (1933a): Beitrag zur Berechnung von Sandfängen. *Ges.-Ing.*, 54, 478.
BLUNK, H. (1933b): Beitrag zur Klärung der Vorgänge bei der biologischen Reinigung von Abwasser durch Tropfkörper. *Ges.-Ing.*, 36—37.
BOGÁRDI, J. (1972): *Sediment Transportation in Streams*, Akadémiai Kiadó, Budapest. (In Hung.)
BÖSS, P. (1955): *Technische Hydromechanik. Taschenbuch für Bauingenieure*. Springer Verlag, Berlin.
BRATBY, J.—MARAIS, G. v. R. (1974): Dissolved air flotation. *Filtration and Separation*, **11**, 6, 614—624.
BRATBY, I.—MARAIS, G. v. R. (1977): Dissolved-air flotation in activated sludge. *Prog. Wat. Tech.*, 9, 311—322.
BULKAI, L. (1973): Experiences with clarifiers applied in water treatment technology. *Vízügyi Közlemények*, 175—198. (In Hung.)
BURDICH, J. (1964): *Pomery proudeni ve vertikalnich dosazovacich nadrzich*. Praha—Podbaba.
BURGESS, F. J.—GILMOUR, C. M.—MERRYFIELD, F.—CARSWELL, J. K. (1961): Evaluation criteria for deep trickling filters. *Journal WPCF*, **33**, 8, 787—799.
CAMP, T. R. (1942a): Grit chamber design. *Sewage Works J.*, **14**, 2, 368.
CAMP, T. R. (1942b): Effect of turbulence in retarding settling. *Proceedings of the Second Hydraulic Conference*, 307.
CAMP, T. R. (1946): Sedimentation and the design of settling tanks. *Trans. Am. Soc. Civ. Engrs*, 111, 895.
CAMP, T. R. (1953): Studies of sedimentation basin design. *Sew. and Ind. Wastes*, **25**, 1, 1.
CAMP, T. R. (1955): Flocculation and flocculation basin. *Trans. ASCE*, 120, 1.
CAMP, T. R. (1964): Theory of water filtration. *Proc. ASCE, J. San. Engrg. Div.* 90, SA4, 1—30.
CAMP, T. R.—STEIN, P. C. (1943): Velocity gradient and internal work in fluid motion. *J. Boston Soc. Civ. Engrs*, 30, 219—237.
CLEASBY, J. L.—BAUMANN, E. R. (1962): Selection of sand filtration rates. *J. Am. Wat. Works. Ass.*, 54, 579—602.
DALLAS, J. L. (1958): The uniflow tank. Parts I—II. *Water and Sewage Works*, May, 210—215; June, 249—254.
DOBBINS, W. E. (1944): Effect of turbulence on sedimentation. *Trans. ASCE*, **190**, 629.
DOBBINS, W. E. (1964): Mechanism of gas absorption by turbulent liquids. *Int. Conf. on Water Poll. Res. 1962*, Pergamon Press, New York.
ECKENFELDER, W. W. Jr.: (1970): *Water Quality Engineering for Practising Engineers*. Barnes and Noble, New York.

ECKENFELDER, W. W.—BARNHART, E. L. (1963): Performance of a high rate trickling filter using selected media. *Journal WPCF*, **35**, 12, 1535—1551.

ECKENFELDER, W. W.—O'CONNOR, D. J. (1961): *Biological Waste Treatment.* Pergamon Press, New York.

EMDE, W. v. d. (1961): Belüftung, Arten und Systeme — Sauerstoffzufuhr und Energieaufwand. *Berichte der ATV*, 13, Die Hamburger Tagung.

EMDE, W. v. d. (1961): Belüftung, Arten und Systeme — Sauerstoffzufuhr und Energieaufwand. *Berichte der ATV*, 13, Die Hamburger Tagung.

EMDE, W. v. d. (1964): Die Technik der Belüftung in Belebtschlammanlagen. *Schweizerische Zeitschrift für Hydrologie*, 26, 338—361.

EMDE, W. v. d. (1969): Belüftungssysteme und Beckenformen. *Vom Wasser.*, Wien.

FAIR, G. H.—GEYER, J. CH.—OKUN, D. A. (1968): *Water and Wastewater Engineering.* Vols I—II. John Wiley and Sons, New York–London–Sydney.

FITCH, E. B. (1966): A mechanism of sedimentation. *Ind. and Engrg. Chem. Fundamentals*, **5**, 1, 129—134.

FORCHEIMER, Ph. (1930): *Hydraulik.* Teuber Verlag, Berlin—Leipzig.

FOX, D. M.—CLEASBY, J. L. (1966): Experimental evaluation of sand filtration theory. *J. of the Eng. Div.*, SA5, 61—82.

FRANZINI, J. B.—HASSAN, N. A. (1964): Hydraulics on thin film flow. *Proc. ASCE*, 90 HY2, 23—36.

GEIGER, H. (1942): Sandfänge für Abwasserkläranlagen. *Archiv für Wasserwirtschaft.*

GOULD, B. W. (1974): Hindered settling and sludge blanket clarifiers. *Effluent and Water Treatment Journal*, **14**, 3, 131—139.

GROCHE, D. (1964): *Die Messung von Fliessvorgängen in ausgeführten Bauwerken der Abwasserreinigung.* Komissionsverlag R. Oldenbourg, München.

GRUHLER, J. F. (1962): Hydraulische Vorgänge in Tropfkörper. *Wasserwirtschaft—Wassertechnik*, **12**, 2, 82—83.

GYÖRGY, I. (1974): *Handbook of Water Engineering.* Műszaki Könyvkiadó, Budapest. (In Hung.)

HALVORSON, H. O. (1936): Aero-filtration of sewage and industrial wastes. *Water Works and Sewerage*, 9, 307—313.

HANEY, P. D. (1956): Principles of flocculation related to water treatment. *Proc. ASCE*, 82 HY4.

HANSON, A. T.—CLEASBY, J. L. (1990): The effects of temperature on turbulent flocculation: fluid dynamics and chemistry. *Journal AWWA*, 11, 56—73.

HARRIS, S. S.—KAUFMANN, W. J.—KRONE, R. B. (1966): Orthokinetic flocculation in water purification. *Proc. ASCE*, 92 SA6, 95—111.

HARTMANN, H. (1966): Der belüftete Sandfang. *Gas- und Wasserfach*, 20, 559—563.

HARTMANN, L. (1967): Influence of turbulence on the activity of bacterial slimes. *Journal WPCF*, **39**, 957—964.

HARTMANN, L.—LAUBENERGER, G. (1968): Influence of turbulence on the activity of activated sludge flocs. *Journal WPCF*, **40**, 4, 670—676.

HAZEN, A. (1892): Some physical properties of sands and gravels. *J. State Board of Health*. 24th Annual Report.

HAZEN A. (1904): On sedimentation. *Trans. Am. Soc. Civ. Engrs*, 53, 63.

HERBERT, D. (1961): A theoretical analysis of continuous culture systems. Continuous culture of micro-organisms. *Chem. Soc. Ind. Monograph*, 13.

HÖRLER, A. (1964): Beitrag zur Steiggeschwindigkeit und zum Steigwiderstand von Luftblasen in reinem Wasser. *GWF*, **105**, 28, 764—765.

HÖRLER, A. (1969): Entwurf von Absetzbecken. Entwurf, Bau und Betrieb von Abwasserreinigungsanlagen. *Wiener Mitteilungen Wasser-Abwasser-Gewässer*, Wien, **4**.

HORVÁTH, I. (1965): Ähnlichkeitsbedingungen für die Untersuchung von Sickerströmungen. *Österreichische Wasserwirtschaft*, **5-6**, 129—135.

HORVÁTH, I. (1966a): Modelling of oxygen transfer processes in aeration tanks. *3rd Int. Conf. on Water Poll. Res.*, Munich.

HORVÁTH, I. (1966b): Hydraulic investigation of sand trap with horizontal rotor. *Vízügyi Közlemények*, 4, 539—544. (In Hung.)

HORVÁTH, I. (1970): Model studies on turbulence in aeration basins. *Res. in Water Quality and Water Technology*, VITUKI, Budapest,

HORVÁTH, I. (1972): The role of recirculation rate in the activated-sludge waste treatment process. *Hidrológiai Közlöny*, 10, 436—440.

HORVÁTH, I. (1974): Some hydraulic problems of ultra-high rate filtration. *First World Filtration Congress*, Paris.

HORVÁTH, I.: (1975). Ultra-high rate filtration problems. *Journal AWWA*, August, 452—453.

HORVÁTH, I. (1984): *Modelling in the Technology of Wastewater Treatment*. Pergamon Press. Oxford–New York–Toronto–Sydney–Paris–Frankfurt.

HORVÁTH, I.—MUSZKALAY, L. (1969): Investigation of settling tanks and determination of some basic data. *Vízügyi Közlemények*, 3, 413—424. (In Hung.)

HOWLAND, W. E. (1958): Flow over porous media as in a trickling filter. *Proceedings, 12th Ind. Wast. Conf.*, Purdue University, 94, 435—444.

HUDSON, H. E. (1965): Physical aspects of flocculation. *Journal AWWA*, 7, 885—892.

IMHOF, K. (1966): *Taschenbuch der Stadtenwässerung*. 21. verbesserte Auflage. Verlag von R. Oldenbourg, München–Wien.

IMHOFF, R.—ALBRECHT, D. (1972): Zum Einfluss von Temperatur und Turbulenz auf dem Sauerstoffeintrag im Wasser. *GWF*, **113**, 6, 264—268.

ISON, C. R.—IVES, K. J. (1969): Removal mechanisms in deep bed filtration. *Chemical Engineering Science*, 24, 717—729.

IVES, K. J. (1968): Theory of operation of sludge blanket clarifiers. *Proc. Instn. Civ. Engrs*, 39, 243—260.

IVES, K. J. (1969): Theory of filtration. Special Subject No. 7. *International Water Supply Congress, Vienna.*

IVES, K. J. (1970): Rapid filtration. *Water Research*, 4, 201—223.

IVES, K. J.—GREGORY, J. (1967): Basic concepts of filtration. *Proc. of the Soc. for Water Treatment and Examination*, 16, 147—169.

IVICSICS, L. (1968): *Hydraulic Models*. Műszaki Könyvkiadó, Budapest. (In Hung.)

IWASAKI, T. (1937): Some notes on sand filtration. *J. Am. Wat. Works Ass.*, 29, 1591—1602.

KAEDING, J. (1962): Beitrag zur Schlammeindickung durch Flotation. *Wasserwirtschaft–Wassertechnik*, **12**, 1, 6.

KALBSKOPF, K. H. (1961): Über den Einfluss von Detergentien auf die Umwälzströmung in Belüftungsbecken. *Technische-Wissenschaftliche Mitteilungen der Emschergenossenschaft und Lipperverbandes*, H. 4, 123.

KALBSKOPF, K. H. (1966a): Über den Absetzvorgang in Sandfängen. *Veröffentlichungen des Institutes für Siedlungswasserwirtschaft der TH Hannover*, H. 24, 78.

KALBSKOPF, K. H. (1966b): Strömungsverhältnisse und Sauerstoffeintragung bei Einsatz von Oberflächenbelüftern. *Vom. Wasser*, 33, 154—171.

KALINSKE, A. A. (1958): *Flotation in Waste Treatment. Biological Treatment of Sewage and Ind. Wastes*. Reinhold Publ. Co., New York.

KALINSKE, A. A. (1970): *Turbulence Diffusivity in Activated Sludge Aeration Basins*. Pergamon Press, New York.

KALINSKE, A. A.—SHELL, G. L.—LASH, L. D. (1968): Hydraulics of mechanical surface aerators. *Water and Wastes Engineering*, 4, 65—68.

KALMAN, L.: (1966): *Beitrag zur Messung von Fliessgeschwindigkeiten in Klärbecken mit Thermosonden*. Basel.

KARELIN, J. A. (1955): Waste-water cleaning in mineral processing factories. *Vodosnabzhenyiye i sanitarnaya teknika*, 2.

KATZ, W. J. (1958): Sewage sludge thickening by flotation. *Public Works*, 89, 114.

KATZ, W. J.—GEINOPOS, A.—MANCINI, J, L. (1962): Concepts of sedimentation applied to design. Part I—II. *Water and Sewage Works*, April, 162—165; May, 169—259.

KAWAMURA, S. (1973): Coagulation considerations. *Journal AWWA*, **65**, 6, 417—423.

KAYSER, R. (1967): Ermittlung der Sauerstoffzufuhr von Abwasserbelüftern unter Betriebsbedingungen. *Veröff. Inst. f. Stadtbauwesen TH Braunscheig*, H. 1.

KIRSCHMER, O. (1926): Untersuchungen über den Gefälleverlust am Rechen. *Mitteilungen des hydraulischen Institutes der TH. München*. 1, 21.

KNOP, E. (1951): Über die Absetzvorgänge in Klärbecken. *Ges.-Ing.*, 9.

KNOP, E. (1952): Über dem Einfluss der Strömung auf Flockung und Absetzvorgänge in Klärbecken. *Ges.-Ing.*, **73**, 9—10, 157—163.

KNOP, E.—BISCHOFSBERGER, W.—STALMANN, V. (1964): *Versuche mit verschiedenen Belüftungssystemen im technischen Masstab*. Teil 1—2. Vulkan Verlag, Essen.

KNOP, E.—KALBSKOPF, K. H. (1968): Energy and hydraulic tests on mechanical aeration systems. *4th Int. Conf. on Water Poll. Res., Prague*, Pergamon Press, New York.

KOVÁCS, GY. (1972a): *Hydraulics of Seepage*. Akadémiai Kiadó, Budapest. (In Hung.)

KOVÁCS, GY. (1972b): *Researches and Experimental Results*. 35, VITUKI, Budapest. (In Hung.)

KRONE, R. B.—ORLOB, G. T.—HODGKINSON, C. (1958): Movement of coliform bacteria through porous media. *SJW*, 1.

KROPF, A. (1957): La sédimentation des matières grenues. Applications: dessableurs et separateur d'huille. *Revue Suisse d'Hydrologie* **19**, 1, 320—342.

KROUPA, V. (1960): Beitrag zur Berechnung von Schnellfiltern. *WWT*, 7, 310—315.

KULSKIY, L. A. (1964): *The Chemistry and Technology of Water Treatment*. Műszaki Könyvkiadó, Budapest. (In Hung.)

KYNCH, G. J. (1952): A theory of sedimentation. *Trans. Faraday Soc.*, 48, 161.

LAUBENBERGER, G. (1970): Struktur und physikalisches Verhalten der Belebtschammflocke. *Karlsruher Berichte*, H. 3.

LESENYEI, J. (1953): *Treatment of Industrial Wastes*. Közlekedési Kiadó, Budapest. (In Hung.)

LEVICS, V. G. (1958): *Physico-chemical Hydrodynamics*. Akadémiai Kiadó, Budapest. (In Hung.)

LEVIEL, R. (Ed.) (1979): *Degremont Water Treatment. Handbook*. Halsted Press, New York, 653.

MACKRLE, S. (1965): Hydrodynamic principles of sludge blanket stability. *Effluent* and *Water Treatment Journal*, **5**, 10, 505—512.

MANZ, H. (1972): Die hydraulische Regelung von Belebtschlammanlagen. *GWF*, **113**, 3, 131—132.

MARSKE, D. M.—BOYLE, J. D. (1973): Chlorine contact chamber design — A field evaluation. *Water and Sewage Works*, January.

McKINNEY, R. E. (1962): Mathematics of complete-mixing activated sludge. *Proc. ASCE, J. San. Eng.*, Div. 88 SA3, 87—113.

MELTZER, D. (1965): An idealized theory of biological filtration efficiency. *Journ. Inst. of Sew. Purif.* 2, 181—182.

MERKEL, W. (1971): Die Bemessung horizontal durchströmter Nachklärbecken von Belebungsanlagen. *GWF*, **112**, 12, 598—600.

MERKEL, W. (1974): Die Bemessung vertikal durchströmter Nachklärbecken von Belebungsanlagen. *GWF*, **111**, 6, 272—277.

METCALF, L.—EDDY, H. P. (1979): *Wastewater Engineering: Treatment Disposal, Reuse*. McGraw Hill Co., New York, 920.

MICHAU, R. (1951): Pressure-diagrams in filters. *L'Eau*, 38, 191—194.

MILBURY, W. F.—PIPES, W. O.—GRIEVES, R. B. (1965): Compartmentalization of Aeration Tanks. *Proc. ASCE, J. San. Eng. Div.*, 92, 45—61.

MINTS, D. M. et al. (1966): Modern theory of filtration. Special Subject, No. 10. *International Water Supply Association*, Barcelona.

MÖLLER, U. (1963): Berechnung der Sauerstoffaufnahme und wirtschaftliche Gestaltung von Druckluft-Belüfungsbecken. *Stuttgarter Berichte zur Siedlungswasserwirtschaft TH Stuttgart*.

MONGAJT, J. L.—RODZILLER, J. D. (1958): *Metodi ochistki stochnikh vod*. Gostoptekhizdat, Moscow.

MONTGOMERY, J. M. (1985): *Water Treatment. Principles and Design*. John Wiley and Sons, New York, 227.

MUELLER, J. A.—VOELKEL, K. G.—BOYLE, W. C. (1966): Nominal diameter of floc related to oxygen transfer. *Environmental San. Eng.*, Div. 92, 9—20.

MÜLLER-NEUHAUS, G. (1952): Über Klärung und Flockung von Abwasser. I—II. *Ges.-Ing.*, 73, 7—8, 132—136; 73, 11—12, 194—198.

MÜLLER-NEUHAUS, G. (1952/53): Über die Kennzeichnung der hydraulischen Verhältnisse bei Klarbecken. *Die Wasserwirtschaft*, 1, 7—10.

MURPHY, K. L.—BOYLER, B. J. (1970): Longitudinal mixing in spiral flow aeration tanks. *Proc. ASCE, J. San. Eng. Div.*, SA2, 211—221.

MURPHY, K. L.—TIMPANY, P. L. (1967): Design and analysis of mixing for an aeration tank. *Proc. ASCE, J. San. Eng. Div.*, 93, SA5, 1—15.

MUSZKALAY, L.—VÁGÁS, I. (1954): Determination of hydraulic efficiency in settling-tanks. *Hidrológiai Közlöny*, 11/12, 461—473. (In Hung.)

NEIGHBOR, J. B.—COOPER, Th. W. (1965): Design and operation criteria for aerated grit chambers. *Water and Sewage Works*, 12, 448—454.

NÉMETH, E. (1963): *Hydromechanics. I.* Tankönyvkiadó, Budapest. (In Hung.)

ÖLLŐS, G. (1970): The role of slow sand filters in water treatment. *Hidrológiai Közlöny*, No. 1, 1—12. (In Hung.)

ORDON, CH. J. (1966): Manifolds, rotating and stationary. *Proc. ASCE, J. San. Eng.*, Div. 92, SA1, 269—280.

OSTERMANN, G. (1967): *Neuere Erkenntnisse über Wirkungsweisen und Leistungen von Tangential- und Quersandfängen.* Dissertation, TH Dresden.

OTTENGRAF, S. P.—RIETEMA, K. (1969): The influence of mixing on the activated sludge process in industrial aeration basins. *Journal WPCF*, 41, 8, R 282—293.

PALLASCH, O.—TRIEBEL, W. (1967): *Lehr- und Handbuch der Abwassertechnik.* Band I—III. Verlag von Wilhelm Ernst and Sohn, Berlin–München.

PALLASCH, O.—TRIEBEL, W. (1985): *Lehr- und Handbuch der Abwassertechnik. I.—IV.* (Dritte Auflage.) Verlag von Wilhelm Ernst und Sohn, Berlin–München.

PASVEER, A. (1956): *Oxygen Supply as a Limiting Factor in Activated Sludge Purification. Biological Treatment of Sewage and Industrial Wastes.* Vol. I. Reinhold Co., New York.

PATÓ, T. (1961): Some questions of activated sludge aeration tanks with horizontal rotors. *Hidrológiai Közlöny*, 6, 481—489. (In Hung.)

PATTANTYÚS, Á. G. (1958): *Applied Hydraulics.* Tankönyvkiadó, Budapest. (In Hung.)

PFLANZ, P. (1966): Über das Absetzen des belebten Schlammes in horizontal durchströmten Nachklär-becken. *Veröff. d. Inst. f. Siedlungswasserwirtschaft der TH. Hannover*, H. 25, Hannover.

PFLANZ, P. (1969): Performance of (activated sludge) secondary sedimentation basins. *4th Int. Conf. IAWPR*, Prague, Pergamon Press.

PÖNNINGER, R. (1965): *Biologische Abwasserreinigung durch Tropfkörper.* Verlag der Österreichischen Abwasserrundschau, Wien.

PÖPEL, F. (1949): Bemessung von Absetzbecken. *Ges.-Ing.*, 70, 15—16, 241—247.

PÖPEL, K. (1978): *Aeration and Gas Transfer.* Delft University of Technology, Delft.

PÖPEL, F. (1943): Die Leistung und Berechnung von Spültropfkörpern. *Beihefte zur Ges.-Ing., II. 21.* R. Oldenbourg Verlag, München.

PÖPEL, F.—HARTMANN, H. (1958): Der neue belüftete Sandfang auf der biologischen Reinigungsanlage der Stadt Heilbronn. *Gas- und Wasserfach*, 22, 535—537.

PÖPEL, F.—WEIDNER, J. (1963): Über einige Einflüsse auf die Klärwirkung von Absetzbecken *GWF*, 104, 28, 795—803.

POPP, P. et. al. (1973): Die physikalisch-chemischen Grundlagen der Wasserbehandlung durch Flockung. *WWT*, 23, 2, 42—49.

RANDOLF, R. (1966): *Kanalization und Abwasserbehandlung.* VEB Verlag für Bauwesen, Berlin.

REBHUN, M.—ARGAMAN, Y. (1965): Evaluation of hydraulic efficiency of sedimentation basins. *J. of the San. Eng.*, Div. 10, 37—45.

318   References

RINCKE, G.—MÖLLER, V. (1966): *Einfluss von Einblastiefe und Luftdurchsatz auf Sauerstoffzufuhr und Sauerstoffertrag bei feinblasiger Breitbandbelüftung.* Westdeutscher Verlag, Köln.
RÖSSERT, R. (1964): *Hydraulik im Wasserbau.* R. Oldenbourg Verlag, München.
ROUSE, H. (1958): *Engineering Hydraulics.* John Wiley and Sons Inc., New York.
RUMPF, H. (1972): Verfahrenstechnische Gesichtspunkte bei der Leichtflüssigkeitabscheidung. *Symposium Leichtflüssigkeitsabscheidung, Bürstadt, Sept. 29.*
RUMPF, A. (n. d.): *Tropfkörper Bauart Passavant.* Selbstverlag Passavant-Werke.
RUMPF, H.—RAASCH, J. (1962): Desagglomeration in Strömungen. *Symposium Zerkleinern.* VDI-Verlag GmbH, Düsseldorf, 151—159.
SAITENMACHER, L. (1965): Der Einfluss der Temperatur auf Absetzanlagen. *Wissenschaftliche Zeitschrift der TH Dresden,* 5, 1247—1954.
SAWYER, C. M.—KING, P. H. (1969): The hydraulic performance of chlorine contact tanks. *Purdue Industrial Waste Conference,* Purdue University.
SCHULTZ, G. (1956): Der Turmtropfkörper des Weisselsterverbandes. *Wasserwirtschaft–Wassertechnik,* 7.
SCHULZE, K. L. (1960): Trickling filter theory. *Water and Sewage Works,* 3, 100—103.
SCHWALDENBACH, W. (1964): Vergleichender Überblick amerikanischer und deutscher Erfahrungen in der maschinellen Ausrüstung mechanischer und biologischer Kläranlagen. *Kommunalwirtschaft,* 9, 397—402.
SEELIGER, R. (1949): Gasblasen in Flüssigkeiten. *Die Naturwissenschaften,* 36, 41.
SIERP, F. (1953): *Die gewerblichen und industriellen Abwassern.* Springer Verlag, Berlin–Göttingen–Heidelberg.
SINKOFF, M. D.—PORGES, R.—McDERMOTT, J. H. (1959): Mean residence time of a liquid in a trickling filter. *Proc. ASCE, J. San. Eng. Div.,* **86,** 6, 51—59.
SMOLUCHOWSKI, M. (1917): Versuch einer mathematischen Theorie der Koagulationskinetik Kolloider Lösungen. *Z. Physik. Chem.,* 92, 129—168.
SONTHEIMER, H. (1965): Verfahrenstechnische Richtlinien für Bau und Betrieb von Flockungsanlagen und für die Koagulationsfiltration. *Veröffentlichungen Wasserchemie,* Karlsruhe, H. 3.
SOUCEK, J.—SINDELAR, J. (1967): *The Use of a Dimensionless Criterion in the Characterization of Flocculation.* Praha–Podbaba.
STEINOUR, R. H. (1944): Rate of sedimentation. *Ind. and Engrg. Chem.,* **36,** 501, 618, 804.
STOBBE, G. (1964): Über das Verhalten von belebten Schlamm in aufsteigender Wasserbewegung. *Veröff. d. Inst. f. Siedlungswasserwirtschaft der TH Hannover,* H. 18, Hannover.
STOKES, G. G. (1845): On the theories of the internal friction of fluides in motion and of the equilibrium and motion of elastic solids. *Transactions, Cambridge Philosophical Soc.,* Vol. VIII., 287—958.
SUSHKA, J. (1971): Oxygenation in aeration tanks. *Journal WPCF,* **43,** 1, 81—92.
SZABÓ, L. (1969): Experiments with plate-type oil separators. *Hidrológiai Közlöny,* 1, 6—15. (In Hung.)
SZALAY, M. (1967): *Biology in Engineering Practice.* Műszaki Könyvkiadó, Budapest. (In Hung.)
TESARIK, J. (1963): Geschwindigkeiten in Flockenwirbelsichten und Aufenthaltszeiten in Schlammkontaktanlagen. *WWT,* **13,** 6, 258—261.
TESARIK, J. (1967): Flow in sludge blanket clarifiers. *Proc. Am. Soc. Civ. Engrs,* 93, SA6, 105—120.
TETTAMANTI, K. (1972): Chemical Operations. I. Tankönyvkiadó, Budapest. (In Hung.)
THIRUMURTHI, D.: (1969): A breakthrough in the tracer studies of sedimentation tanks. *Journal WPCF,* **41,** 11, 405—410.
TRAWINSZKI, H. (1953): Hydrocyklon als Hilfsgerät zur Grundstoff-Veredelung. *Chemie-Ing. Techn.,* 6, 331—341.
VAN VUUREN, L. R. J.—STANDER, G. J.—HERZEN, M. R.—VAN BLERK, S. V. H.—HAMMAN, P. F. (1968): Dispersed air flocculation flotation for stripping of organic pollutants from effluents. *Water Research,* 2, 177—183.
VELIKANOV, M. A. (1955): Dinamika ruslovih potokov. Gostekhizdat, Leningrad. — Mixing and sedimentation basins. *Journal AWWA,* 47, 8, 768—790.

VILLEMONTE, J. R.—ROHLICH, G. A.—VALLANCE, A. T. (1966): Hydraulic and removal efficiencies in sedimentation basins. *Third Int. Conf. WPCF*, Section II, No. 16, Munich.

VRABLIK, E. R. (1959): Fundamental principles of dissolved-air flotation of industrial wastes. *Proc. of the 14th Purdue Ind. Waste Conf.*, 743.

WALKER, J. D. (1968): High energy flocculation and air-and-water backwashing. *Journal AWWA*, 3, 321—330.

WECHMANN, H. (1966): *Hydraulik. 3.* Verbesserte Auflage. VEB Verlag für Bauwesen, Berlin.

WEISS, H. G.—SIEWERT, W. H. (1965): *Die Druck- und Geschwindigkeitsverteilung in einem Hydrocyklon.* Firma Fscher Wyss GmbH, Ravensburg.

WIEDERHOLD, W. (1954): Offene Fragen der Filtertechnik, I—II. *GWF*, X. 658—664; XI. 719—725.

WILSON, F. (1981): *Design Calculations in Wastewater Treatment.* E. and F. N. Spon, London, New York, 221.

WOLF, D.—RESNIK, W. (1963): Residence time distribution in real systems. *Ind. and Eng. Chem. Fundamentals*, **2**, 4, 287—293.

YAO, K. M. (1970): Theoretical study of high-rate sedimentation. *Journal WPCF*, **42**, 2, 218—228.

ZAHRADKA, V. (1967): The role of aeration in the activated sludge process. *Proc. 3rd Int. Conf. on Water Poll. Res.* Washington.

ZUNKER, F. (1938): Theorie des Fettabscheiders und ihre praktische Anwendung. *Ges.-Ing.*, 61, 454—458.